3rd International Academic Conference on Frontiers in Social Sciences and Management Innovation (IAFSM 2020)

Held online due to COVID-19

SHS Web of Conferences Volume 96 (2021)

Beijing, China
20 December 2020

Editor:

Sang-Bing Tsai

ISBN: 978-1-7138-2727-6

Printed from e-media with permission by:

Curran Associates, Inc.
57 Morehouse Lane
Red Hook, NY 12571

Some format issues inherent in the e-media version may also appear in this print version.

No changes have been made to the content of these proceedings. There may be changes to pagination and minor adjustments for aesthetics.

Printed with permission by Curran Associates, Inc. (2021)

For additional information, please contact EDP Sciences – Web of Conferences at the address below.

EDP Sciences – Web of Conferences
17, Avenue du Hoggar
Parc d'Activité de Courtabœuf
BP 112
F-91944 Les Ulis Cedex A
France

Phone: +33 (0) 1 69 18 75 75
Fax: +33 (0) 1 69 28 84 91

contact-edps@webofconferences.org

Additional copies of this publication are available from:

Curran Associates, Inc.
57 Morehouse Lane
Red Hook, NY 12571 USA
Phone: 845-758-0400
Fax: 845-758-2633
Email: curran@proceedings.com
Web: www.proceedings.com

TABLE OF CONTENTS

Preface

Due to recent pandemic, the 3rd International AcademicConference on Frontiers in Social Sciences and Management Innovation (IAFSM 2020) which was planned to be held in Beijing, China, was held virtually online during December 20, 2020. The decision to hold the virtual conference was made in compliance with many restrictions and regulations that were imposed by countries around the globe. Such restrictions were made to minimize the risk of people contracting or spreading the COVID-19 through physical contact. There were 38 individuals who attended this on-line conference, represented many countries including UAS, Canada, UK, Italy, Thailand and China.

On behalf of the organizing committee of the IAFSM 2020, it is an honor and delight to welcome all participants to attend this conference. This conference has been organized byChongqing University of Posts and Telecommunications. Two main topics have been discussed, i.e. Social Science and Economic Science. We are perceived confidently that this conference will provide positive influence and contribute to develop the academic field. Supporting three main topics, the iafsm 2020 invitedfive experts in the fields. Prof Masami Yoshida, he used to work in the Faculty of Education, Toyama University, and did studies about preservice trainings. He is a member of the Japan Society for Educational Technology, Japan Association for Educational Media Society, Japan Society for Information and Systems in Education, and Japan Association for College and University Education.

The Proceeding of the iafsm 2020,consists of 86 paper submissions, amount 34 papers were accepted by IAFSM 2020. Each of the papers submitted in ees was reviewed by at least two experts using the double-blind system. The published papers have passed all necessary improvement requirements in accordance to the EDP Sciences standard. Similarity tests by Turnitin program (with the highest threshold of 15 %). Last but not least, I personally would like to thank you the official committees, organizing partners, andscientific&editorial board. Special thanks as well to our co-host partners, Central South University University of Ferrara, Lincoln University, University of Texas. I would like to express my gratitude feeling for your participations, and please prepare yourself to gain the treasure of knowledge from the passionate experts. It is my pleasure to see many of you in iafsm 2020 and see you again in iafsm 2021. safe and stay healthy in COVID-19 pandemic.

Sang-BingTsai, China
17th, Dec, 2020

IAFSM2020

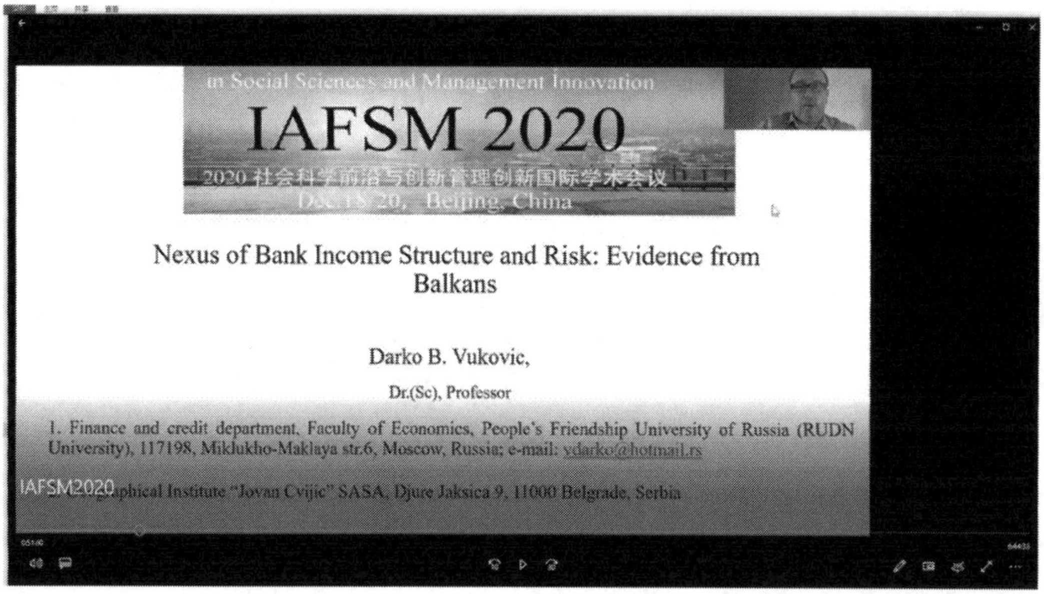
in Social Sciences and Management Innovation
IAFSM 2020
2020 社会科学前沿与创新管理创新国际学术会议
Dec.18-20, Beijing, China
Nexus of Bank Income Structure and Risk: Evidence from Balkans
Darko B. Vukovic,
Dr.(Sc), Professor
1. Finance and credit department, Faculty of Economics, People's Friendship University of Russia (RUDN University), 117198, Miklukho-Maklaya str.6, Moscow, Russia; e-mail: vdarko@hotmail.rs
IAFSM2020
Institute "Jovan Cvijic" SASA, Djure Jaksica 9, 11000 Belgrade, Serbia

2020
3rd International Academic Conference on
Frontiers in Social Sciences and Management
IAFSM2020

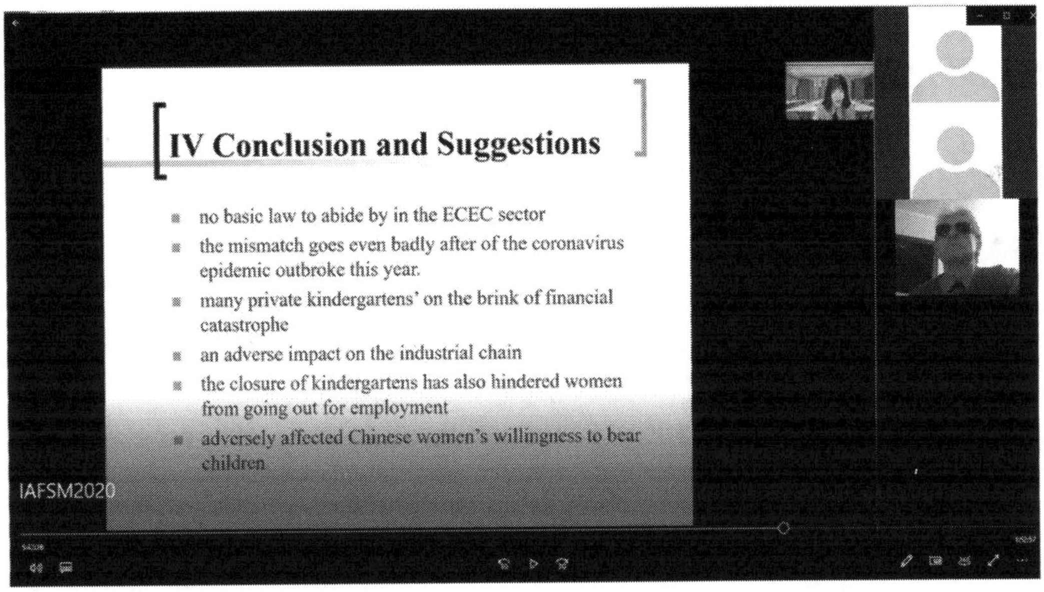
IV Conclusion and Suggestions
no basic law to abide by in the ECEC sector
the mismatch goes even badly after of the coronavirus epidemic outbroke this year.
many private kindergartens' on the brink of financial catastrophe
an adverse impact on the industrial chain
the closure of kindergartens has also hindered women from going out for employment
adversely affected Chinese women's willingness to bear children.
IAFSM2020

Statement of Peer review

In submitting conference proceedings to SHS Web of Conferences, I certify to the Publisher that I adhere to the **Policy on Publishing Integrity** of the journal in order to safeguard good scientific practice in publishing.

1. All articles have been subjected to peer review administered by the proceedings editors.
2. Reviews have been conducted by expert referees, who have been requested to provide unbiased and constructive comments aimed, whenever possible, at improving the work.
3. Proceedings editors have taken all reasonable steps to ensure the quality of the materials they publish and their decision to accept or reject a paper for publication has been based only on the merits of the work and the relevance to the journal.

Title, date and place of the conference

2020 3rd International Academic Conference onFrontiers in Social Sciences and Management Innovation

20th, Dec, 2020 Zoom(Webniar), China

Proceedings editor(s):

Sang-BingTsai

Date and editor's signature

20th, Jan, 2021

Based on the Public Quality Perception Satisfaction Evaluation Theory System Research and Yunnan Province Application Practice

Lei Feng*, *Juxiu* Huang, *Jingxing* Liao

China National Institute of standardization, Beijing 100191, China

Abstract. Survey on public satisfaction of quality work is one of the effective methods for connecting the quality management department with the public. In this paper, the application of techniques and methods of survey on public satisfaction of quality work is studied and explored from the aspects of the design of survey content index system, data collection, and statistical result analysis based on the survey on public satisfaction of quality work in Yunnan Province, so as to provide enlightenment and reference for the relevant research work in the field of quality.
Keywords: Quality work, Satisfaction, Empirical study.

1 Introduction

Today, countries of the world attach great importance to the quality construction and take the quality development as an important tool to maintain their international competitive edges. Survey on public satisfaction of quality work is one of the effective ways to connect the quality management department and the public [1]. Evaluation on the public satisfaction of quality work, on the one hand, pays attention to the comparison process and results of public psychology and learns the quality work of government based on the public perceptions of various satisfaction indexes it; on the other hand, can identify the problems in quality work of local government by systematic data analysis, with precise suggestions put forward, thereby promoting the government to improve the quality work and provide the public with better and effective quality work services.

In this paper, the application of techniques and methods of survey on public satisfaction of quality work is studied and explored from the aspects of the design of survey content index system, data collection, and statistical result analysis based on the survey on public satisfaction of quality work in Yunnan Province (2018-2019), so as to provide enlightenment and reference for the relevant research work in the field of quality [2-3].

2 Research methods

The evaluation on the public satisfaction of quality work is mainly carried out based on product quality, engineering quality, service quality, environmental quality and quality

* Corresponding author: 1053038950@qq.com

awareness. The evaluation index system is built with the public as the core following the evaluation criterion of public perception [4].

2.1 Design of survey content index system

The survey on satisfaction of quality work takes product quality, engineering quality, service quality, environmental quality and quality awareness as the first-level indicators [5]. In this survey, the contents easy to be felt and closely related to daily life are selected the evaluation indexes. The satisfaction survey is a process of quantitative analysis, which is mainly used to reflect the attitude of the surveyed on survey contents with figures. In this study, 5 point Likert scales are adopted for measurement; the specific index system and measurement method are shown in Table 1.

Table 1. The indexes of public evaluation on quality work of the government.

First-level indexes	Measurement indexes	Measurement methods
Product quality	Foods and Drugs	
	Agricultural products such as grain, vegetables, fruits, meat, and aquatic products	
	Consumer goods	
	Special equipment	
Engineering quality	Constructional engineering	
	Public transportation engineering	
Service quality	Short-distance public transport service	
	Long-distance transport service	
	Communication and network service	Very dissatisfied i=1;
	Banking service	
	Insurance service	Dissatisfied i=2;
	E-commerce service	General i=3;
	Logistics and express service	Satisfied i=4;
	Primary and secondary education service	Very satisfied i=5
	Medical service	
	Pension service	
	Public service	
	Tourism service	
	Public cultural and sports services	
	Home renovation service	
	Water quality of rivers and lakes	

Environmental quality	Air
Quality awareness	Quality complaint satisfaction
	Notification of quality safety issues and quality publicity
	Improvement of product, engineering, environment and service quality

2.2 Data collection

The survey adopts the stratified sampling method, which collects data through online survey combined with offline interception visits. This survey recovers 3,700 effective samples of public satisfaction of quality work of Yunnan Province.

2.3 Evaluation method

The satisfaction indexes can reflect the overall feeling and evaluation of the surveyed on a certain industry. The higher average score means the higher satisfaction of the surveyed, and vice versa. The computational formula is shown below:

$$Q = \frac{\sum_{i=1}^{r} m_i q_i}{\sum_{i=1}^{r} q_i} \quad (r=5、4、3、2、1) \tag{1}$$

$$S = \sum_{k=1}^{l} \beta_k \sum_{j=1}^{k} \alpha_j \cdot Q \tag{2}$$

Where, S refers to the total score of government quality work satisfaction; Q refers to the score of satisfaction of a certain measurement index; mi refers to the valuation of satisfaction measurement option of index i (m1=20, m2=60, m3=60, m4=80, m5=100); α j refers to the weight of measurement index j; and β k refers to the weight of the first-level index k. The weight of each index is determined according to the historical survey data and Delphi method.

3 Analysis on applications in Yunnan province

3.1 Statistical results

As shown by the survey results, the comprehensive score of the survey on satisfaction of quality work in Yunnan Province is 69.36, which means that the residents are fairly satisfied with the quality work [6]. As for the first-level indexes, environmental quality is outstanding, with a score of 71.52; followed by engineering quality (70.20), service quality (68.96) and quality awareness (68.15); and product quality has a lower score (67.96), as shown in Fig. 1.

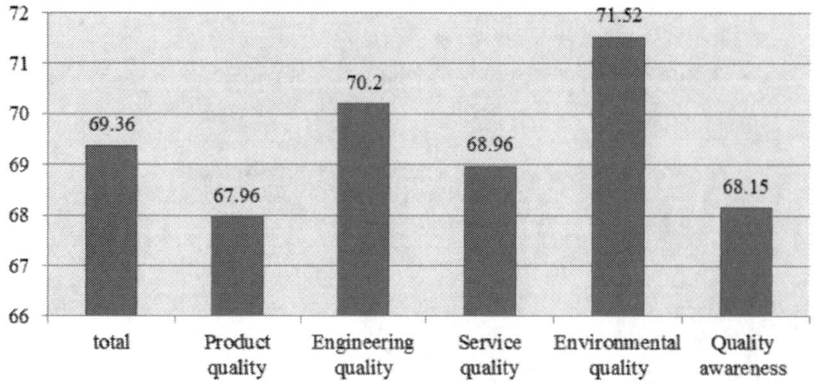

Fig. 1. Overall situation of public satisfaction of quality work in Yunnan Province.

The results of evaluation on product quality, engineering quality, service quality, environmental quality and quality awareness are shown in Fig. 2.

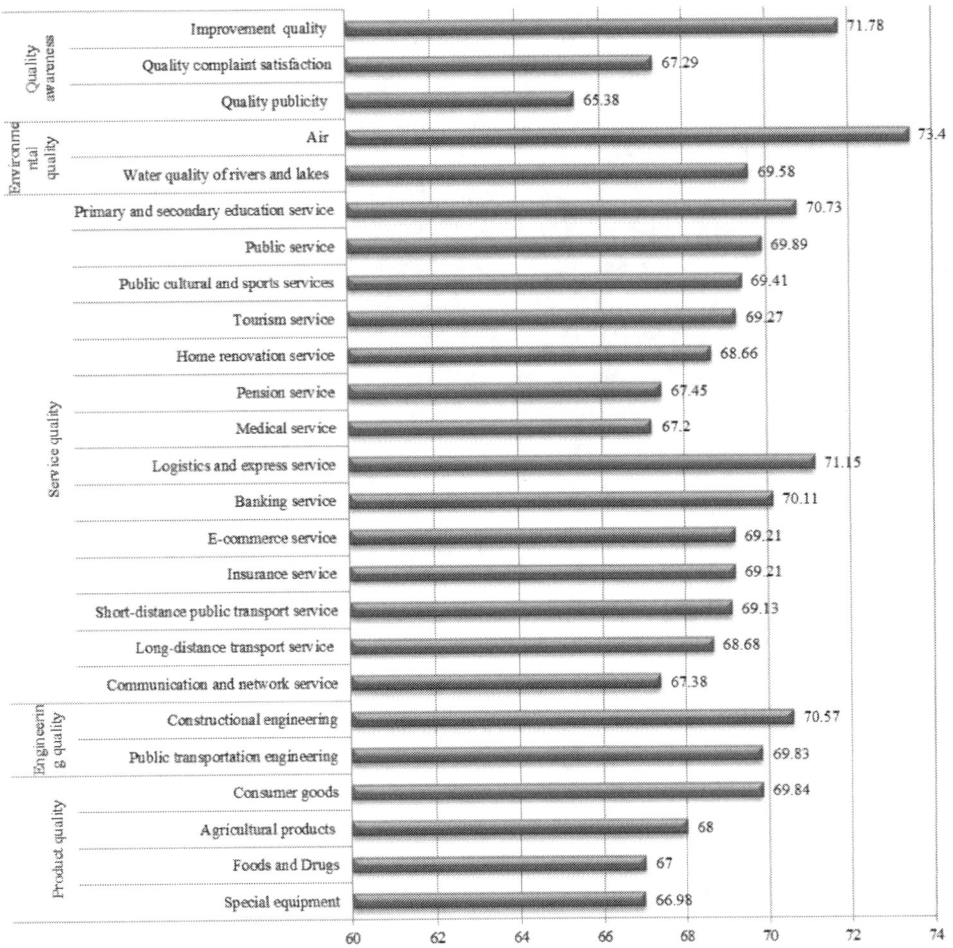

Fig. 2. Results of evaluation on public satisfaction of quality work in Yunnan Province.

3.2 Analysis of influence factors

The essence of quality work satisfaction is the function of difference between the quality effect that can be perceived by the public and the expectation of quality [7]. Therefore, the in-depth understanding of characteristics of the population involved in satisfaction evaluation will provide support for the government to make decisions, and improve the public satisfaction of quality work. Considering that the public expectation may be affected by multiple factors, such as social background, perception, individual demand, previous experience and living area environment, the survey on the satisfaction of quality work shall also be carried out in combination with the above-mentioned variables.

3.2.1 Analysis on the surveyed at different ages

The results show that the score of satisfaction increases with the increase of age of the surveyed. In specific, the ones after 60s (50-59 years old) have the highest satisfaction of government quality work, with the score of 71.19; followed by the ones after 80s (30-39 years old), the ones after 50s (above 60 years old), and the ones after 70s (40-49 years old), with the scores of 70.86, 70.74, and 68.97, respectively; and the ones after 90s (18-29 years old) have the relatively low satisfaction, with the score of 67.23, as shown in Fig. 3.

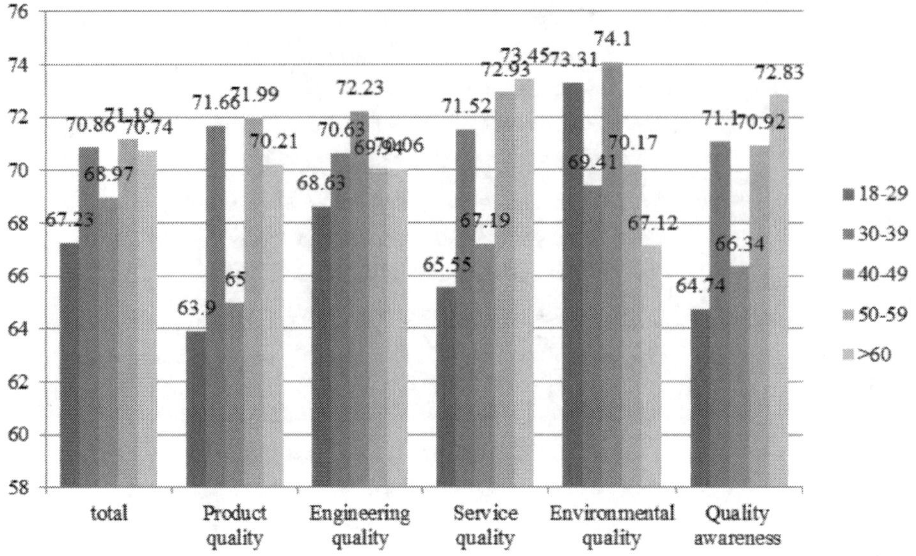

Fig. 3. The satisfaction of government quality work of residents at different ages.

3.2.2 Analysis on the surveyed of different genders

The results show that there is only a small difference in the satisfaction scores of each index between genders; in specific, the scores given by the male are slightly higher than those given by the female. In general, the satisfaction score of the male is 69.71 and that of the female is 69.03. Therefore, the male residents have a higher satisfaction score of quality work of Yunnan Province than the female residents, as shown in Fig. 4.

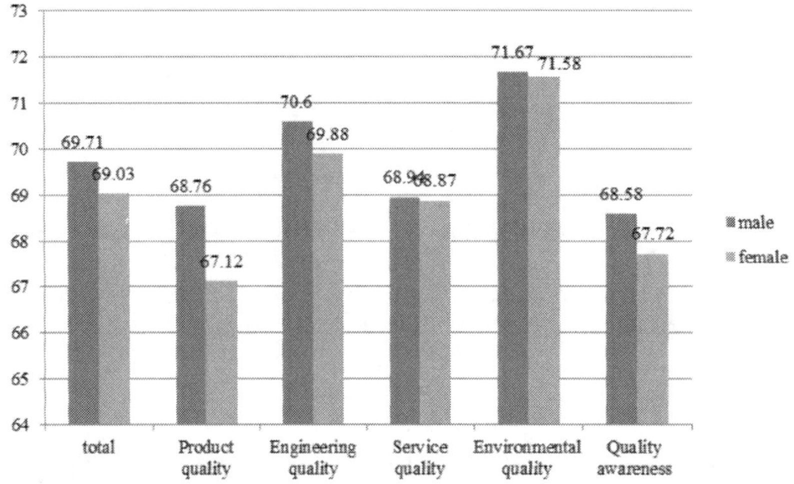

Fig. 4. The satisfaction of government quality work of residents of different genders.

3.3 Analysis on the surveyed with different levels of education

The results show that the score of satisfaction increases with the increase of academic qualifications. The scores of satisfaction are respectively 68.65, 69.39, 69.45, and 69.72 in the population at primary school level and below, population at senior high school/ technical secondary school/technical school level, population at junior high school level, and population at the undergraduate level. The residents at the junior college level have the highest satisfaction of government quality work, with the score of 69.77. The scores of satisfaction of population with different academic qualifications are shown in Fig. 5.

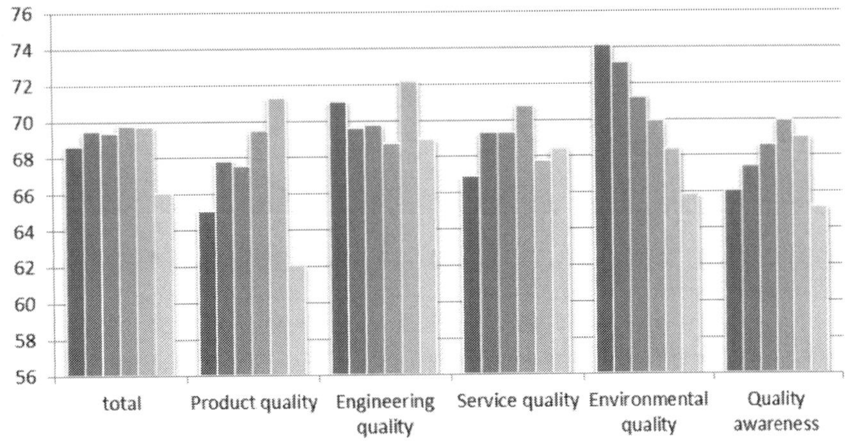

Fig. 5. The satisfaction of government quality work of residents with different levels of education.

4 Conclusion

Based on the empirical study on public satisfaction of government quality work, we have built a quality work satisfaction survey index system. Through collecting the data of survey on the satisfaction of quality work of Yunnan Provincial Government and combining with the statistical findings, we hereby propose the following suggestions for Yunnan Provincial Government to carry out quality improvement actions, strengthen the overall quality supervision, comprehensively improve the quality level, and promote the implementation of provincial construction based on quality:

4.1 Precisely improve quality services aiming at quality shortcomings

The government shall, based on the results of evaluation on the public quality satisfaction survey report, attach importance to quality shortcomings, carry out quality work in a purposeful and focused manner, and improve the satisfaction of the public on quality work. The government shall also improve the complaint handling mechanism, implement an accountability and time-limited response system, and assign a receiver for specific complaint information; the receiver shall be responsible for handling the complaint, and provide result feedback within the specified time, so as to ensure that each complaint can receive a response, thus improving the satisfaction of handling of quality complaints [8].

4.2 Strengthen quality publicity, and help the quality to improve steadily

In order to steadily increase the quality and promote the quality satisfaction of the public, it is suggested that the local governments shall conscientiously implement the decisions and arrangements of the Party Central Committee and the State Council, to improve the development quality and efficiency, and carry out the quality improvement actions focused on the strategy of developing cities (counties) based on quality [9]. At the same time, local governments shall actively carry out all-round publicity and education activities in various forms and contents, innovate in publicity forms of quality work, carry out publicity and education by multiple channels and methods, and cultivate quality awareness and create quality atmosphere in the society. They shall also give full play to the role of news media as the main channel of quality publicity, and adhere to the correct direction of public opinions [10].

4.3 Strengthen quality supervision and promote quality brand building

It is suggested that the local governments shall establish the strong awareness of quality first, carry out quality improvement actions, and vigorously promote standardized production of foods, brand creation, quality and safety supervision; implement the access mechanism of foods and drugs, effectively improve the quality of foods, drugs and agricultural products, continuously improve the product quality, and improve the public satisfaction of product quality. The local governments shall also, based on the national "Internet +" action plan, strengthen the information construction of quality inspection, and gradually release the energy of "Internet + Quality Inspection"; strengthen the effective connection between risk monitoring and random checking; establish a quality safety inspection linkage working mechanism; and precisely, effectively and severely crack down on quality violations, to effectively improve the quality level.

5 Acknowledgements

This research was financially supported by the government quality monitoring and evaluation system improvement and implementation project (582019C-7110).

References

1. M L Morris, R Tripp, and A A Dankyi. Adoption and Impacts of Improved Maize Production Technology: A Case Study of the Ghana Grains Development Project, International Maize and Wheat Improvement Center (CIMMYT), (1999)
2. J O Ouma, F M Murithi, W M wang, H Verkuijl, M Gethi and H D Groote. Adoption of Maize Seed and Fertilizer Technologies in Embu District, International Maize and Wheat Improvement Center (CIMMYT), (2002)
3. F H Zheng, Red Data Book of Evaluation on Performance of Local Governments in China (2012), Xinhua Publishing House, **185**, (2012)
4. D L Wang, X Z Song, Survey and Empirical Study on Public Service Satisfaction – Taking the Survey on Public Industry in Jinan City as an Example [J]. Chinese Public Administration, **6**, (2009)

5. G Casella, R L Berger. Statistical Inference (2nd ed.) [M]. Pacific Grove: Duxbury Press, (2002)
6. D Freedman. Statistical Models: Theory and Practice (revised ed.) [M]. New York: Cambridge University Press, (2009)
7. Q J Shi. Government Performance Evaluation: Concepts, Methods and Applications of Evaluation Results [M]. Beijing: Peking University Press, (2016)
8. P Jin. Study on the "High Quality Development" Economics [J]. China Industrial Economics, **4**, 5-18, (2018)
9. C Z Yang, F Yang. Analysis of Housing Changes in Contemporary Tibetan Families-Based on Census Data [J]. China Tibetology (Beijing), **4**, 103-114, (2018)
10. S Z Zheng, X Zhao. Who can promote the development of high quality: the government or the efficient market? - Empirical Test Based on Panel Threshold Model [J]. Contemporary Economic Management, **42**, (2020)

Research on Promoting the Construction of Online Government Service Platform with Standardization

Xiao Wen XU* *Li* Ying

China National Institute of Standardization, No. 4 Zhichun Road, Beijing, China

Abstract. The standardization of the online government service platform is of great significance to simplify the procedures and improve the efficiency of administrative licensing agencies. In this paper, the definition of online government service platform, from the law, policy background, the practice level, carry out the work of government services online analysis summarizes the necessity, the role of online government service platform and the existing problems, and analysis the reasons from the dimension of standardization, standardization can be targeted to summarize proposed online government service platform construction of three aspects of the role.
Keywords: Research, Online Government Services Platforms, standardization

1 Introduction

An online government services platform is a vetting administration platform that integrates application, data exchange, approval, supervision, inquiry and other services, providing companies and the public with one-stop service in an efficient, transparent and standardized way. Totally different from traditional vetting administration mode, it is based on modern Internet and data technologies, creating an innovative workplace. It demonstrates the restructuring and optimization of vetting structure and process and removes temporal, spatial and division restrictions.

Article 33 of Administrative Licensing Law of the People's Republic of China states that "administrations shall establish and improve related systems, introduce E-government practices, post licensing items on their websites and enable the submission of e-application for administrative licensing".

Since 2015, the State Council issued a series of documents, which focused on the reform of administrative licensing system as an offensive move for the transformation of government functions and a central cannon for administration streamlining and power delegating. In Jan. 2017, Notification of Issuing and Distributing Guidelines for Building 'Internet + Government Service' Technologies System by General Office of the State Council (Guo Ban Han (2016) No. 108) (hereinafter referred to as No. 108) was distributed

* Corresponding author: xuxw@cnis.ac.cn， 420655022@qq.com

to address such issues as an incomplete set of services, insufficient information sharing, low rate of acceptance and inconvenient services for companies and the public;Notification of Sharing General Information on Citizens, Enterprises, Institutions and Social Organizations Contained in Administrative Licenses by State Commission Office of Public Sectors Reform (Administrative Examination and Approval Reform Office of the State Council), National Development and Reform Commission and Ministry of Public Security issued and distributed in Feb. 2017 and Implementation Programme of Integrating and Sharing Information on Government Affairs issued and distributed in May 2017 specifies information sharing for online government services; The State Council's Advices on Quality Improvement Move issued and distributed in Sep. 5, 2017 states that we shall " improve social governance and public service, introduce 'Internet + Government Service', speed up the standardization of administrative licensing system, optimize work flow, streamline administrative process and increase efficiency" ; Passed at the first meeting of the Central Reform Commission presided by Chairman Xi Jinping in March 2018, Advices on Furthering Convenient Vetting Services hold a principle of " integrating institutional innovation and 'Internet + Government Service'" . At the videophone conference on deepening the reform of streamlining administration, delegating powers and improving regulation and services to transform government functions on June 28, 2018, Prime Minister Li Keqiang instructed that we shall " enable companies and the public to go through administrative procedures without asking for a favor, continue with less licenses and convenient services move, reinforce clean-up, reduction and merging, develop 'Internet + Government Services' and facilitate the integration of online and offline operation" .

In conclusion, the mentioned documents above by those agencies reveal the necessity and urgency of building online government services platforms.

2 Application

In recent years, a few local governments have gained experience from persistent, innovative efforts. Zhejiang Province introduced the practice of At Most Once to allow data to be shared rather than being isolated, improve "Internet + Government Services" design at top level and build a comprehensive government service network across Zhejiang using methodology of systems engineering. It also introduced "Non-face-to-face Vetting" and "Online Procedures", which integrate all government service networks of 65 agencies under Zhejiang government and at metropolitan and county level into a single government service network and thus allow information systems for government affairs to be interconnected. In June 2017, Jiangsu Government Service website was launched to ensure that all applications submitted for approval at provincial, metropolitan and county level are addressed as reasonable as possible. By the end of 2017, majority of applications to administrations at provincial, metropolitan and county level had been addressed online, a change from face-to-face to key-to-key. Wuhan in Hubei Province introduced the practice of Right Now, Online and Just Once. In response to the mandate of enabling online operation to be normality and offline one to be an exception, it gave priority to online operation and built a unique ID certification system and a government-specific big data center across Wuhan. In conclusion, these practices reveal that administrative licensing agencies preferred online government services platforms in streamlining process and increasing efficiency.

3 Role

3.1 Less cost and higher efficiency is an important move for a saving and efficient government

Online government services platforms help reduce bureaucratic procedures frequently occurring offline, get rid of disadvantages of inefficient administration such as excessive paper work and meetings and lower government's operating costs. They have changed the situation of each vetting agency operating on its own way into sharing information and cooperating with each other. With vetting procedures optimized and rebuilt, they served the purpose of shortening timeline、cost saving and increasing efficiency

for example, Agriculture in China new veterinary drug registration for examination and approval matters need to submit each evaluation 30 sets of paper material, material printing costs 10000 yuan, agricultural Agriculture in China integrated office system of administrative examination and approval and the veterinary drug evaluation system has realized the paperless review after connected, saved a review of 30 sets of tens of thousands of yuan for the print paper material, new veterinary drug registration for examination and approval matters in accordance with the transferred to an average of 120 items a year, every year for the enterprise to save material printing costs an average of 1.2 million yuan of above.

3.2 Integrating information and resources and publicizing government affairs is a path to a transparent and fair government

Online government services platforms enable transparent procedures for administrative licensing all around, from directory of items under administrative licensing, terms and conditions and procedures to licensing process and measures, which allow licensing-related activities to be supervised by applicants and the public for open and transparent purpose.

3.3 Improving technical regulations and administrating according to the law is a warrant for an honest government based on the rule of law

Licensing procedures and legal documents as stipulated in the Administrative Licensing Law and Guidelines for Standardization of Administrative Licensing may be established via online government services platforms. The Administrative licensing agencies have to operate accordingly. Administrative activities violating these procedures are significantly reduced. With accountability, authority, responsibility, and timelines identified, the licensing process is standardized. Additionally, applicable conditions as stipulated in related laws and regulations, including documents, timelines and fees relating to administrative licensing, may be embedded, so as to reduce omission and illegal act and alert errors.

4 Issues about online government services platforms and their causes from the perspective of standardization

4.1 Rushing forward, operating on its way, duplication and serious waste

At the national level, the top-level design scheme for the overall construction of the government service platform has been issued, and the mandatory requirements such as the time limit and process of the administrative license work have been continuously increased. However, the foundation of the government service platform construction in different regions is inconsistent, and the national top-level design scheme cannot accurately guide the specific practice in different regions. local administrations have to move existing offline operation to online one without any change in mode of operation and an online government services platform in one area may not affect those in other areas. As a result, duplication occurs.

Causes are that unified planning and reasonable deployment is not made at early stage of online government services platform building. Rushing into building these platforms leads to waste and duplication.

4.2 Imbalance, "best game no one played" and even eye candy

Different schedules and design whereby individual online government services platforms were built, different materials required for the same subject matter and different deadlines caused difficulties in coordinated licensing and insufficient licensing via online government services platforms.

Causes are that independent operation and fragmentation are impairing the power of online government services platforms. Even in the same urban district and for the same item, it is impossible to unify item name, materials to be submitted and process and coordinate information or information systems, which is one of causes that online government services platforms are the best game no one played.

4.3 Simple services

Incomplete functionality and simple services lead to online acceptance, vetting, decision, payment and delivery being not realized but provision of general information and receipt of applications. These platforms do not serve the purpose of integrating online and offline operation as expected. Applicants still have to go through online and offline procedures for the same subject matter.

Causes are that it is undesirable to move offline administrative licensing as a whole to online one. What is feasible for offline licensing may not be the case for online one, for example, trans-department and cross-level licensing.

5 Role of standardization in building online government services platforms

5.1 Essential foundation and support

The starting point of setting standards is to "obtain the best order", "promote the best common benefit" and "promote standardization by standardization". The basic system tool of standards can also play its role in online administrative licensing.

5.1.1 Existing deliveries of administrative licensing standardization are applied to online government services platforms to make online government services in line with compliance requirement

Existing deliveries of administrative licensing standardization are applied to online government services platforms to make online government services in line with compliance requirement. Like offline government services, online ones shall be delivered in accordance with general requirements of Guidelines for Standardization of Administrative Licensing (2016) and No. 108. Deliveries of offline government service standardization such as items, work flow and services relating to administrative licensing are applied to online government services to make them in line with these documents. For instance, such provisions as "acceptance forms or receipts shall be issued within 5 days after applications for administrative license are submitted" and "decisions about administrative licensing shall be posted online according to the law" may be embedded into the work flow of online government services to make these services in line with compliance requirements.

5.1.2 Elements and work flow for an item are standardized for coordination purpose

Items are identified according to the rule of Minimum Granularity. The name, justification, code, deadline, required materials, qualification for acceptance, qualification for approval, procedures, results document name of any item are standardized based on the principle of The Same Item, Standard, Code and Process, Minimum Granularity. This standardizing move lays the groundwork for the integrated vetting on online government services platforms and in turn enables Three-level Linkage or Five-level linkage between administrative licensing at provincial, metropolitan, county, town and village level. It eventually leads to "any application, even from area other than your own one, can be accepted via the nearest platform in the same city."

5.2 Higher service quality

Give full play to the forward-looking guidance function of standards in online administrative licensing to help standardize the management of administrative licensing items and reduce and optimize the process.

5.2.1 Apply the standardization research results to the construction of online government service platform

Administrative Licensing Items Classification and Coding Rules under research specify rules and methodology for coding administrative licensing items. These rules ensure that each item has a unique identification, which facilitates not only administrative licensing management and implementation but also administrative licensing-related information systems building. Instead of separating information, they allow data statistics, comparison, analysis and application in response to needs for interconnection between information systems and resources sharing.

5.2.2 Results under research are applied to online government services platform building to guide the optimization of online government services process

It is undesirable to move offline administrative licensing apply mechanically online one. The former shall be refined in line with online scenario. Administrative Licensing Process Optimization Methodology & specifications under development provide guidelines for identifying, renovating and refining online administrative government service progress.

5.3 Guidance on evaluation

Performance requirements for online administrative licensing are developed to guide online government services platform building. Provisions of No. 108 and compliance requirements of Guidelines for Standardization of Administrative Licensing dealing with administrative licensing may be integrated into these performance requirements.

5.3.1 Standardizing online government services platform

Performance requirements for online administrative licensing enable administrations to plan and deploy online government services platforms in a unified way, identify roadmap to growth, eliminate independent operation, duplication and imbalance and lay the foundation for online administrative licensing interconnection and information sharing.

5.3.2 Efficient online services. Performance requirements for online administrative licensing provide self-inspection and inspection guidelines for online government services platforms

They facilitate tracking and evaluating performance online, identifying and correcting problems and refining online government services platforms and their processes.

6 Acknowledgements

This article in this phase is part of a state-funded key research program, namely, Research on standardization of Press, Publication, Radio, Film and Television Administrative License work (No.572018Z-6569).

References

1. J H Li, Analysis of Issues from Online Administrative Vetting and Measures in China, Tianjin's Economics, **12**, 35-37 (2013)
2. Ouyang Keli, Study on Chinese Online Administrative Vetting Building: Example of Area Q [D], Public Administration School, Nanchang University, Jiangxi
3. H Qin, Study on Reform of Chinese Administrative Vetting Mode: Based on Practices of Administrative Service Centre [D], Jilin University, Jilin 133-137(2011)
4. X Han, Thinking and Advices on Standardization of Guangdong Administrative Licensing Item Vetting [J], Standard Science, **5** (2017)
5. SN Ying. Contemporary Chinese Administrative Law (Vol)[M]. Beijing: Founder Press, China ,**735** (2005)
6. C T Li. General introduction to standardization (6th edition) [M]. Beijing: Renmin University Press, **48**(2014)
7. Unger, law in modern society, translated by wu yuzhang and zhou hanhua, China university of political science and law press, p 43 (1994)
8. H B Long, X J Wang. Dividends and performance evaluation of administrative approval reform: 2013-2015. China development press, p 98 (2016)

Obstacles to the Development of BIM in China's Prefabricated Construction: A Decision-making Test and Evaluation Laboratory (DEMATEL) Model

Yanan Ma, *Qiong* Shen*, and *Lei Tian*

College of Architecture and Urban-Rural Planning, Sichuan Agricultural University, Dujiangyan Sichuan, 611830, China

Abstract. With the continuous progress of the time, technology is also developing rapidly, among which *Building Information Modeling (BIM)* has also become a hot topic in prefabricated constructions. Previous research has made it clear that BIM can promote the development of prefabricated construction. However, in practical applications, BIM technology has not been well developed in prefabricated construction. In order to find out the obstacles to the development of BIM technology in China's prefabricated construction, and identify the relationship between obstacles, this study identified 20 obstacles through literature review. This study used the *Decision-making Test and Evaluation Laboratory (DEMATEL)* model to get the importance of each obstacle and the causality between them, and provides a solution to the development of BIM technology in China to promote the development of BIM technology in new construction methods.
Keywords: BIM, Prefabricated construction, Obstacles.

1 Introduction

In the 1980s, with the proposal of building energy efficiency, green buildings are gradually emerging in China. The traditional cast-in-place building model consumes more energy than green building, and it is also likely to have a greater impact on the surrounding environment [1], which does not meet China's requirements for sustainable and green development. The appearance of prefabricated construction conforms to the requirements of green building, which embodies the advantages of green building. It has the advantages of energy-saving, environmental protection, material saving, reduction of labour and so on [2]. In order to face the urgent needs of all parts of the country to transform and upgrade to the modernization of the construction industry, the state and industry have successively issued relevant guidelines and policies. Since 2018, China's prefabricated construction market has grown rapidly. According to national policy directives, it is estimated that the national prefabricated construction market will have a space of about 2 trillion by 2020. Newly assembled buildings accounted for 15% of new buildings, and further reached 30% in 2025. With the improvement

* Corresponding author: swydong@sicau.edu.cn

of technology and management level, prefabricated construction will have a broad market and space in China.

In the context of the continuous progress of the times, BIM technology has gradually become familiar with China's construction industry. It can help various professions in the construction industry to implement the integration of construction information. So it can improve the construction efficiency of prefabricated buildings, which is beneficial to development [3]. Because the development of BIM conforms to the principle that the more buildings are built off site and then onsite, the more labor and material costs are saved [4]. There are many advantages to using BIM technology in prefabricated construction. However, according to the perspective of development status, the situation of BIM in China is not optimistic. BIM is facing not only domestic application in prefabricated construction, but also obstacles to its implementation [5]. Scholars such as Oesterreich and Hong separately analyzed the development of BIM and the obstacles to the implementation of prefabricated construction [6, 7]. Some scholars have linked the two together. For example, Tan analyzed the relationship between BIM's obstacles to the development of prefabricated construction [8]. However, there are not enough obstacles involved in these studies. Due to the influence of analytical methods, some factors that are considered not important would be screened out. Therefore, the purpose of this study is to expand the number of obstacles based on previous research, and use DEMATEL model to find the most important obstacles. Then analyzed the causal relationship between obstacles. According to the results, this study analyzed the obstacles to the development of BIM in China's prefabricated buildings and gave appropriate suggestions.

2 Obstacles to BIM applications in prefabricated construction

Based on the development status of BIM in prefabricated construction, it is necessary to analyze the obstacles to the development of BIM in prefabricated construction. In this study, the literature review method was used to check, compare, analyze and sort out the related obstacles. And on this basis, the existing unhealthy industrial phenomena are improved, and the obstacles summarized are made into expert questionnaires. Consultation and interviews were conducted with experts who have been engaged in related research and work for more than 3 years, and the obstacles were modified and supplemented. Finally, 20 obstacles were obtained, which are representative of the BIM implementation in prefabricated buildings, as showed in Table 1.

Table 1. Obstacles are affecting development of BIM in prefabricated construction.

Number	Obstacles	Related references
F1	High precision requirements for fabricated components	[8]
F2	Lack of motivation and promotion	[8]
F3	Lack of insurance applicable to BIM implementation	[8]
F4	Lack of intellectual property protection	[8]
F5	New construction methods take a long time to learn	[8]
F6	High initial costs	[9]
F7	Lack of uniform industry guidance standards	[9]
F8	The construction industry is dispersed	[9]
F9	Lack of standards to quantify BIM benefits	[9]
F10	Conservative industry culture and resistance from client	[9]
F11	Stakeholders' negative treatment	[10]
F12	Lack of experienced practitioners	[10]
F13	Lack of collaborative work processes	[10]
F14	Lack of professional interaction	[10]
F15	Lack of standard contract templates	[10]
F16	Resistance to data sharing	[10]
F17	Compatibility of BIM software	[11]
F18	Lack of research on BIM implementation	[11]
F19	Lack of BIM experts	[11]
F20	Equipment purchase and maintenance costs	[12]

3 Research methods

Methods used in this study is the Decision-making Test and Evaluation Laboratory (DEMATEL). The DEMATEL method can calculate the influence degree and the affected degree of each element on other elements through the relationship between the elements in the system and the direct influence matrix. Then calculate the cause degree and the center degree of each element according to the previous step as the basis for constructing the model. So that it can determine the causal relationship between the elements and the position of each element in the system. According to this model, the main obstacles of BIM in prefabricated construction could be obtained and then draw influence. The steps of the research method are as follows:

Step 1: Through literature review, sort out and select 20 influencing factors that hinder the development of BIM in prefabricated construction, then make a questionnaire.

Step 2: This study uses the questionnaire survey method to quantify the relationships between the various obstacles, the former factor is considered to have an influence on the latter factor, and the "1" is considered to have no influence, and the "0" is considered to have no influence. The data was collected by distributing 60 questionnaires to personnel of construction units, managers, structural designers, scientific research institutes and other personnel engaged in the construction industry. A total of 44 valid questionnaires were collected in this survey, with an effective recovery rate of 73%.

Table 2. Division of direct influence between obstacles.

Proportion	0	1-25	26-50	51-75	76-100
Score	0	1	2	3	4

Step 3: The questionnaire data are collected. Then the impact degree is divided according to the proportion of interviewees who think Fi has an impact on Fj (see Table 2).According to the scale of 0 ~ 4, the direct influence degree between the obstacles is divided into 5 levels: none, small, normal, large and larger (the corresponding scores are 0, 1, 2, 3, 4). Through the data of the statistical questionnaire, the direct influence matrix F is obtained (see Table 3).

Step 4: Normalization is a routine operation to standardize things. So it needs to be calculated to the direct impact matrix normalization. The normalization matrix X is obtained by the following formula:

$$X = (\frac{f_{ij}}{Max})_{n\times n} \quad (i \text{ and } j \in 1,2,3,\Lambda ,n). \tag{1}$$

Table 3. Direct influence matrix F.

	F1	F2	F3	F4	F5	F6	F7	F8	F9	F10	F11	F12	F13	F14	F15	F16	F17	F18	F19	F20
F1	0	0	0	0	2	3	2	0	0	3	3	2	2	0	0	0	0	0	2	2
F2	0	0	0	0	2	0	3	0	0	3	4	4	2	2	2	2	1	2	2	0
F3	0	0	0	0	0	3	1	0	0	3	3	0	0	0	0	0	0	0	0	2
F4	0	0	0	0	0	0	2	1	2	2	0	0	0	2	2	3	0	2	2	0
F5	0	0	0	0	0	3	0	2	2	3	3	3	2	2	2	2	1	3	3	0
F6	1	1	1	0	2	0	2	3	1	3	3	3	1	1	0	2	2	2	2	3
F7	0	0	0	0	3	2	0	2	3	3	2	3	3	2	2	1	1	2	2	0
F8	0	0	2	0	3	3	3	0	1	4	3	2	2	2	2	2	1	2	1	2
F9	0	0	0	0	0	0	2	0	0	3	3	0	2	2	0	0	0	2	3	0
F10	0	0	1	0	2	2	2	2	2	0	3	2	2	3	2	3	2	2	2	1
F11	0	0	0	0	2	1	2	2	2	3	0	3	2	4	0	3	0	3	2	0
F12	1	0	0	0	0	3	3	2	1	3	2	0	2	2	2	2	2	2	2	0
F13	0	0	0	0	3	1	2	2	2	3	2	2	0	3	2	0	2	3	3	0
F14	0	2	1	0	3	1	3	4	3	3	2	3	3	0	2	3	2	3	3	1
F15	0	0	1	0	0	0	3	2	3	3	2	0	2	0	0	0	0	3	0	0
F16	0	0	1	0	2	1	3	2	2	2	2	2	3	3	2	0	2	2	2	1
F17	0	0	0	0	3	2	2	2	0	3	2	2	3	2	0	0	0	3	2	0
F18	1	2	1	1	3	2	2	2	3	3	2	2	4	4	2	2	3	0	3	1
F19	0	0	1	0	3	3	3	2	3	2	2	3	4	4	2	2	3	3	0	1
F20	0	0	1	0	0	4	0	1	0	2	2	0	0	0	0	1	0	0	0	0

Step 5: The next step is to calculate the total relation matrix *T*:

$$T = (N + N^2 + N^3 + \Lambda + N^K) = \sum_{k=1}^{\infty} N^k \rightarrow T = N(I - N)^{-1}. \tag{2}$$

Step 6: According to the total relation matrix *T*, the influence degree and the affected degree of each element are further calculated. The influence degree is recorded as *Di* and the affected degree is recorded as *Ci*, and the following formula is obtained :

$$D_i = \sum_{j=1}^{n} t_{ij}, (i = 1,2,3,\Lambda\ ,\ n);\ \ C_i = \sum_{j=1}^{n} t_{ji}, (i = 1,2,3,\Lambda\ ,\ n)\ . \tag{3}$$

Step 7: The center degree *Mi* and the cause degree *Ri* are calculated as following formulas:

$$M_i = D_i + C_i;\ \ R_i = D_i - C_i\ . \tag{4}$$

4 Research result

4.1 Center degree analysis

According to the calculation of the data analysis software, the results are shown in Table 4. The data in Table 4 are formed in Fig.1. From the cause-result graph, the top five factors of center degree are F10, F18, F14, F19 and F11. Center degree indicates the position of the factor in the evaluation index system and its importance in the problem. Therefore, from the results of the center degree, it can be concluded that these five factors have greatly hindered the development of BIM technology in prefabricated construction. Among them, the conservative industry culture and resistance from client has the greatest impact, which reflects that the construction industry should currently be opened, and related customers should also support emerging technologies.

Table 4. Comprehensive influence relationship of obstacles.

Obstacles	Di	Ci	Mi	Ri	Obstacles	Di	Ci	Mi	Ri
F1	0.8994	0.1425	1.0419	0.7570	F11	1.3238	1.8362	3.1600	-0.5124
F2	1.2870	0.2460	1.5330	1.0410	F12	1.2780	1.5731	2.8511	-0.2952
F3	0.4946	0.4502	0.9448	0.0444	F13	1.3408	1.7255	3.0663	-0.3847
F4	0.7988	0.0503	0.8491	0.7485	F14	1.7977	1.7231	3.5208	0.0747
F5	1.3777	1.4758	2.8536	-0.0981	F15	0.8139	1.0589	1.8728	-0.2450
F6	1.3942	1.4059	2.8001	-0.0116	F16	1.3771	1.2308	2.6079	0.1462
F7	1.3517	1.6649	3.0166	-0.3132	F17	1.1971	1.0436	2.2407	0.1534
F8	1.4735	1.4241	2.8976	0.0494	F18	1.8236	1.7153	3.5389	0.1083
F9	0.7988	1.3817	2.1805	-0.5828	F19	1.7592	1.5607	3.3199	0.1986
F10	1.4200	2.2008	3.6208	-0.7809	F20	0.4705	0.5682	1.0387	-0.0976

4.2 Cause degree analysis

If the cause degree is greater than 0, it indicates that the factor has a greater influence than other factors, which is called the cause factor. Otherwise, it is called the result factor. The top five factors with positive cause degree are F2, F1, F4, F19 and F17, indicating that the lack of motivation and promotion has the greatest impact on other factors. The results also show that "Lack of intellectual property protection", "High precision requirements for fabricated components", "Lack of BIM experts", "Compatibility of BIM software" have medium influence on other factors. For example, a certain profit needs to be obtained in an assembly project for stakeholders. However, BIM technology and prefabricated construction are in the early stages of integration. The relevant national policies are not yet perfect, and stakeholders cannot see profits, which will make it more difficult for the market to adapt to the application of BIM technology.

The top five factors with negative cause degree are F10, F9, F11, F13 and F7. The results show that these factors will be greatly affected by other factors. From the above rankings, it can be concluded that the conservative industry culture and resistance from client is the most susceptible to other factors.

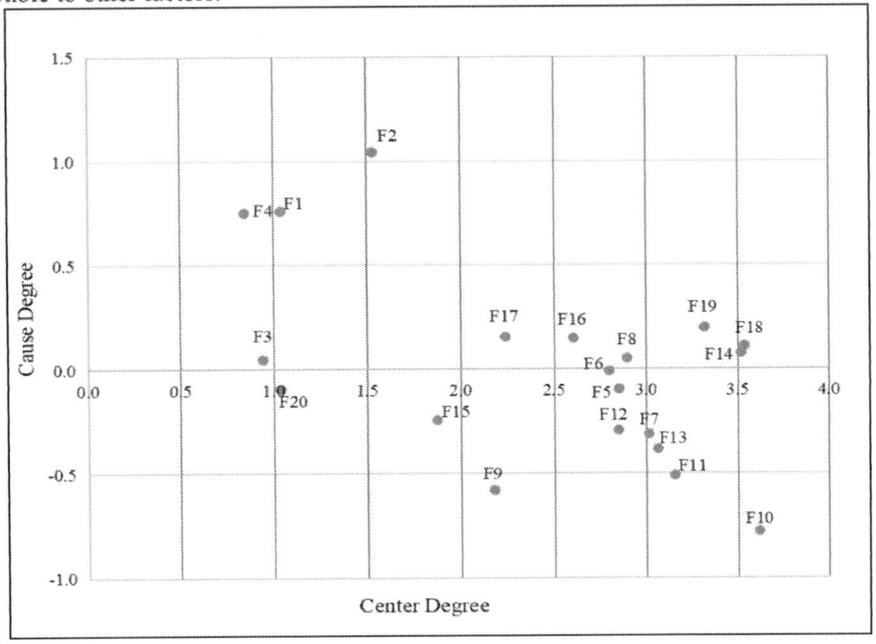

Fig.1. Cause-result graph.

4.3 Discussion

From the above calculation results, it can be seen that the conservative industry culture and resistance from client most likely to be affected by other factors, and this factor has the greatest obstacle to the development of BIM technology in prefabricated construction. And the biggest impact on other obstacles is the lack of corresponding incentives and promotion in China. Therefore, before solving the development of BIM technology in prefabricated construction, this factor should be solved first. To solve the problem of lack of corresponding incentives and promotion, the government should not simply formulate relevant incentive policies. But it needs to be practiced or simulated to calculate the cost of using BIM technology in prefabricated construction, and compare it with the cost of not using BIM technology. Find out where the increased costs of using BIM technology are then formulate targeted incentive and promotion policies for the increased costs. This will not only compensate for the loss of stakeholders using BIM technology in prefabricated construction, but also ensure that stakeholders can obtain a certain profit in it, so as to achieve the purpose of incentive promotion. Effective national policies will gradually open up the industry culture, and customers will gradually accept it. When the primary problem is solved, the secondary problem will be alleviated, so the impact of other obstacles will be reduced accordingly, and BIM technology will be more improved in prefabricated construction.

5 Conclusion

Based on the previous research, this study uses the DEMATEL model to expand the obstacles to twenty. According to the calculation results of the model, first of all, the conservative industry culture and resistance from client has the greatest hindrance to the development of BIM technology in prefabricated construction, and is most susceptible to other factors. Secondly, the lack of research on BIM implementation, the lack of professional interaction, the lack of BIM experts, the negative treatment of stakeholders, and other factors also have hindered the development of BIM technology in prefabricated construction. Thirdly, the lack of incentives and promotion has the greatest impact on other factors. Therefore, the following suggestions can be adopted to appropriately reduce the obstacles to the development of BIM technology in China's prefabricated construction:

(1) Relevant policies formulated by the government for incentives and promotion can be more targeted to achieve the purpose of effectively making up for the losses of stakeholders, and allowing stakeholders to obtain certain profits by using BIM technology;

(2) The relevant laws and regulations should be improved to protect the intellectual property rights emerging technologies. The government should strengthen coordination and cooperation among departments, and hold more high-quality academic conference, in order to increase relevant study and interaction of BIM technology;

(3) The government can increase its encouragement and publicity to promote the cooperation between enterprises and universities to train BIM professionals in order to increase the enthusiasm of stakeholders.

Although this study is helpful for the further development of BIM technology in prefabricated construction, there are still two limitations. The first is that the DEMATEL method cannot rank the degree of obstruction of each factor, so it can only provide a general solution, and cannot provide detailed solution steps based on the problem. The second is that the scope of the questionnaire survey is only in Chengdu, the number of questionnaires in this study is insufficient. Therefore, future research directions should focus on major cities in China, expand the number of questionnaires and the scope of investigations, and consider adopting more comprehensive analytical methods to obtain more accurate conclusions.

References

1. X. Cao, X. Li, Y. Zhu, Z.J.J.o.c.p. Zhang, *A comparative study of environmental performance between prefabricated and traditional residential buildings in China*, J. Clean. Prod. **109** 131-143 (2015).
2. T.U. Ganiron, *Development and Efficiency of Prefabricated Building Components*, Int. J. Smart Home, **10** 85-94 (2016).
3. M. Oh, J. Lee, S.W. Hong, Y.J.A.i.c. Jeong, *Integrated system for BIM-based collaborative design*, AUTOMAT CONSTR, **58** 196-206 (2015).
4. V. Jadhav, T. Nagwanshi, Y. Patil, D.J.v.-. Patil, IRJET, 2076-2078 (2017).
5. B. LIU, W.J.C.E. LIU, *Application status and obstacles research of BIM in domestic construction industry*, J. Constr. Econ., **9** (2015).
6. T.D. Oesterreich, F. Teuteberg, *Behind the scenes: Understanding the socio-technical barriers to BIM adoption through the theoretical lens of information systems research*, J.Technol. Forecast. Soc. Chang., **146** 413-431 (2019).
7. J. Hong, G.Q. Shen, Z. Li, B. Zhang, W.J.J.o.c.p. Zhang, *Barriers to promoting prefabricated construction in China: A cost–benefit analysis*, J. Clean. Prod. ,**172** 649-660 (2018).

8. T. Tan, K. Chen, F. Xue, W.J.J.o.C.P. Lu, B*arriers to Building Information Modeling (BIM) implementation in China's prefabricated construction: An interpretive structural modeling (ISM) approach*, J. Clean. Prod. ,**219** 949-959 (2019).
9. F. Abanda, J. Tah, F.J.J.o.B.E. Cheung, *BIM in off-site manufacturing for buildings*, J. Build. Eng., **14** 89-102 (2017).
10. [10] S. Azhar, M. Khalfan, T.J.C.E. Maqsood, *Building information modelling (BIM): now and beyond*, AJCEB, **12** 15-28 (2012).
11. [11] J. Zhang, Y. Long, S. Lv, Y.J.P.e. Xiang, *BIM-enabled modular and industrialized construction in China*, Procedia Eng., **145** 1456-1461 (2016).
12. [12] Y.-C. Lin, Y.-C.J.T.S.W.J. Su, *Developing mobile-and BIM-based integrated visual facility maintenance management system*, Sci. World J., **2013** (2013).

The Influence and Countermeasures of Citizen Participation in Crisis Management in the Internet Age

Ke YAO*

No.5 Guang Ming Road, Xi'an, Shanxi Province, University of Defense Technology, P.R. China

Abstract. Citizen participation mainly focuses on citizens 'participation in social policies and supervision of public power. Citizen participation in the Internet age has shown some new characteristics in terms of scope of participation, willingness to participate, and means of participation, and it has a profound impact on the process, environment and standards of social crisis management. Public administration departments are urgently required to respond appropriately in terms of regulations, expression channels, etc.
Keywords: Citizen participation, Crisis management, Cyber age

1 The definition and characteristics of citizen participation in the cyber age

Crisis is usually a combination of a variety of unfavorable situations, such as serious threats or challenges to the core values of policy makers, insufficient information, highly uncertain developments and the need for rapid decision-making [2]. Crisis management, as an organized, planned and continuous dynamic management process, is a series of control actions taken by the government or organization at different stages of crisis development in response to potential or current crises in order to effectively prevent, deal with and eliminate crisis [3]. Participation is one of the core concepts of political theory. Citizen participation originally emphasized the confirmation and maintenance of the national political system. With the rise of governance theory and public management theory, the meaning of citizen participation has gradually expanded and deepened, and it has begun to focus on social policy participation and public power supervision [4]. Based on the above understanding, citizen participation in the Internet era mainly refers to the process in which citizens participate in the management of public affairs by means of information under the network background and exert influence on the formation of public policies.

Citizen participation in the Internet age has some typical characteristics. Firstly, the breadth and depth of citizen participation has been expanded. The characteristics of open network, virtual, instant, and sharing have greatly expanded the breadth and depth of citizen participation in public affairs. Citizens who are willing can participate in public affairs at anytime, anywhere, intentionally or unintentionally. The distinctive and segmented participation model has been hit and challenged. But it is worth noting that online hotspots

* Corresponding author: 463872349@qq.com

change frequently, and the degree of citizen participation is constrained by the continuity of the issues and the intensity of the overlay. Secondly, the ability and willingness of citizens to participate has been significantly enhanced. With the widespread use of the Internet, the "information gap" between public power managers and ordinary people in the incident was quickly filled. Ordinary people can grasp information at the same time or even earlier, so as to intervene managers' decisions by the advantage of information acquisition. On the one hand, it has led to a marked increase in the willingness of citizens to participate. On the other hand, it also puts forward higher requirements on the ability of ordinary citizens to discern and to deal with public affairs. Thirdly, the methods and means of citizen participation are more diverse. Ordinary citizens in the Internet age can express their opinions and influence decision-making through network comments, and can also organize network collective activities and network signatures to conduct network mobilization to form consensus and raise issues. In addition, the rapid development of information technology has made it possible to use big data and other means to analyze and collect citizens 'propensity for certain issues and related activity information, which means that the problem of citizens' passive participation has become increasingly prominent.

2 Analysis of the impact of citizen participation on crisis management in the Cyber age

2.1 Accelerate the evolution of the crisis management process

Citizen participation in the cyber age has exposed the crisis management process to more external controls. The typical crisis management process generally includes prevention, emergency, recovery and other stages. The operation of each stage is in a relatively closed environment. There is a clear sequence between the stages. People are also used to mobilizing resources in the process of stylization, coordinate information and solve problems. In the Internet age, all aspects of crisis management are in a more transparent and open state, and are always subject to public supervision and feedback. The public has turned from a service object to a participant and the evolution of crisis management has changed from the previous single thread to multi thread. For example, the new coronavirus crisis management that broke out at the beginning of the year, under the deployment of a game of chess across the country, has a distinction between prevention, emergency, and recovery in different regions. At the same time, there has also been a situation where community prevention, medical department treatment, and public service system gradually resumes production go hand in hand.

2.2 The crisis management environment is becoming more and more complex

Crisis management in the Internet age faces a more complex management environment. The first is that management entities are more diversified. The main body of public crisis management in the cyber age is showing a diversified trend. In the traditional sense, crisis management and control mainly rely on administrative and organizational forces. With the development of mobile terminals and social media platforms in the Internet era, individuals and temporary groups are more widely involved in crisis prevention, evaluation, and recovery. The process of crisis management is facing scientific and refined challenges. Secondly, the crisis management environment is more complex and changeable. The network can not only contain the crisis, but also create it. The Internet has greatly promoted the communication and exchange between people, essentially changed the way of communication between organizations, society and individuals, and enhanced people's ability to manage and control crises. But at the same time, the Internet has quietly lowered the aggregation standards of social relations and speech circles, accelerated the evolution of issues, and made rational information in crisis management interfered by emotional information, attitude information, and moral judgment, and related behaviors are difficult to define, which may cause the crisis to be amplified and superimposed, and weaken people's ability to manage the crisis.

2.3 Crisis management standards are more diverse

Crisis management aims to solve the crisis and reduce disasters, and management effectiveness is an important criterion for judging its success. However, with the development of the Internet and the popularization of online citizen participation, moral factors are becoming an important consideration in judging crisis management activities. On the one hand, the network has the powerful function of aggregating and releasing public opinion. The process of releasing public opinion is not only a process of forming public opinion, but also a process of generating social judgment. This process inevitably leads to the emergence of moral judgment. On the other hand, to more followers, they can also feedback to the crisis manager more quickly and force them to respond to it, resulting in moral judgment becoming another important indicator of crisis management activities.

3 Suggestions on improving citizens' participation in crisis management in the cyber age

3.1 Actively adapt to environmental changes

Firstly, Crisis managers in the Internet age must actively update traditional thinking, adapt to the new public opinion environment, and use new media to respond to the crisis with the assistance of ordinary people. Secondly, transition from the focus on information and flow to the focus on "people" and “relationships” can’t be ignored. We can reduce the social resistance of emergency management by guiding public opinion and responding to concerns. At last, it is necessary to use the more credible platform to release information in a timely manner to strengthen the guidance of public opinion, pay attention to using the power of “action”, and take the initiative to increase the right to speak and influence.

3.2 Strengthen the construction of expression channels

Although China's civic participation has developed in the Internet era, it is still in a relatively autonomous and disorderly stage, and has not yet formed a joint force to become the mainstream. In crisis management, it is necessary to strengthen the construction of channels for citizen participation, such as publicizing emergency management plans on the Internet, collecting improved suggestions, organizing management education activities, etc. Additionally, we can inform citizens about relevant content in advance, increase the transparency of policy decisions and encourage Citizens of all classes, ages, and genders to participate in it and enhance the fairness of decision-making.

3.3 Improve the construction of laws and regulations

At present, the efficient mobilization of the whole society to actively participate in crisis management activities is not only a social development requirement in the information age, but also a realistic choice for us to respond to crises. Facing the new situation of citizen participation, we should gradually improve the construction of laws and regulations on citizen participation and strengthen the protection of online citizen participation and reward and punishment of relevant responsibilities by standardizing the procedures of citizen participation, defining powers and obligations, and strengthening information release management.

References

1. G Plancque, D You, E Blanchard, V Mertens, C Lamouroux. *Role of chemistry in the phenomena occurring in nuclear power plants circuits*, in Proceedings of the International Congress on Advances in Nuclear power Plants, ICAPP, 2-5 Nice, France (2011)
2. W Wang. Citizen Participation. Renmin University of China Press, 2009.
3. X Zhang, J L Wang, 25 Years of Chinese Internet. The official account of People's Network, (2019)
4. L Xue, Q Zhang, K H Zhong. Challenges facing China during the crisis management transition period. Beijing: Tsinghua University Press, 41-45, (2003)
5. C F Zhang. Public Crisis Management: Fully Integrated Model and China's Strategic Choice. China Administration, **77**, (2003)
6. L Ping. Moral Analysis of Citizens' Daily Behavior. Beijing: People's Publishing House, **111** (2004)

Challenges and Countermeasures of China's Cyberspace Governance in the New Era

Fengyun Zhao*, *Yaning* Shi, and *Ke* Yao

National University of Defense Technology, Xi'an, China

Abstract. In the context of the new era, countries are constantly competing and cooperating in the field of cyberspace. China is facing an increasingly complex international environment. Meanwhile, the number of domestic cyber citizens is increasing and the structure is diversified, the incidence of cyber misconduct is increasing significantly. However, policies and laws for cyberspace governance in China are relatively imperfect. Facing with these challenges, China can carry out the practice of cyberspace governance from three aspects: pay attention to internet technology innovation, promote the governance model with multi-actors participation, and improve institutional guarantee.

Keywords: Cyberspace, Cyberspace governance, Cyberspace security.

1 Introduction

The situation in the cyberspace field has shown many changes in the new era. On the one hand, the international cybersecurity situation has changed abruptly, cyber security, development and freedom are the common goals pursued by all the countries Chuanying Lu, 2016 [1], and a community with a shared future has become the appeal of the times Zhengzhong Xu, 2020 [2]. The advent of the 5G era led to a deeper contest between China and the United States in the field of cyberspace. Cyber weapons, hacker and other cyber threats to cyber security are increasing. On the other hand, China's demand for Internet application and security surges, and the development of digital economy and industry advances by leaps and bounds. Digital civilization depends on cyberspace, the rapid development of Internet of Things catalyzes the update and iteration of internet technology. The proportion of the industrial economy built on the network platform is increasing, the internet popularity rate of the public is rising constantly. Hence, the security management and comprehensive governance of the cyberspace are facing unprecedented challenges. Cyberspace governance has entered a new era.

* Fengyun Zhao: fengyunzhao320@163.com

2 Cyberspace and cyberspace governance

2.1 The concept and evolution of cyberspace

The term cyberspace first appeared in the short sci-fi novel by Canadian writer William Gibson to portray the unreal space experienced by computer enthusiasts. With the development of information technology and human knowledge, the connotation of cyberspace has been continuously enriched.

In 2003, the US National Strategy for Cybersecurity identified cyberspace as a space made up of thousands of interconnected computers, servers, routers, switches and optical cables that enable critical infrastructure to function properly. The International Telecommunication Union (ITU) has defined cyberspace as a physical or non-physical domain consisting of all or part of elements include"ng computers and their systems, networks and their software, data and users. "Human interaction" and "exchange of information" are highlighted in the definition of cyberspace in the British Cyber Security Strategy published in 2011. Therefore, cyberspace includes communication service networks, military networks, industrial networks, agricultural networks, etc., which connects physical space and continuously blends with physical space. Cyberspace is becoming a tool for understanding and transforming physical space. Since the establishment of the ARPA network during World War II in the United States, cyberspace has gone through the communication network stage, the information network stage, and the computing network stage Hongren Zhou, 2019 [3]. The first stage is based on basic communication needs. While in the second stage, network is not just a carrier of information, but also an operating platform for various businesses, shopping, social networking, teaching, finance, war, can all be conducted through the Internet, with "Internet +"as the main symbol. In the third stage, informatization moves towards socialization and molecularization, big data, artificial intelligence, global Internet of Things, cloud computing, have become the four major technologies of the new generation of informatization. Based on the four major systems, a global integrated super-large system driven by artificial intelligence is built. In the future, the direction of cyberspace evolution is the cyber-physical space. Cyberspace and physical space are more closely connected and eventually become a whole.

2.2 Connotation of cyberspace governance

Cyberspace governance refers to the international coordination and cooperation activities carried out by all stakeholders in the international community (national governments, private sectors, civil society and even users) in order to promote the orderly and healthy development of cyberspace based on the dual impact of technology and society on the development of the global Internet. Cyberspace governance can be divided into two areas: national governance and global governance, both of which are indispensable. Facing cyberspace governance, on the one hand, we should grasp the basic structure of cyberspace governance in China, on the other hand, we should pay attention to the current mode of global cyberspace governance.

The cyberspace governance structure is based on the three-tier architecture of Internet governance proposed by the Internet Corporation for Assigned Names and Numbers (ICANN), namely, the infrastructure layer, logical layer, economic and social layer Baoguo Cui, 2020 [4]. The infrastructure layer is mainly composed of internet switching center, terrestrial cables, submarine cables, satellites, wireless systems. The logic layer mainly solves the problems of root servers, domain name, IP address, and protocol parameters. The economic and social layer belongs to the content application layer, which is used for information dissemination and citizens' social and service rights.

Two patterns of global governance have gradually evolved over time in the course of global cyberspace governance practice: multilateralism and multistakeholderism. The former focuses on national sovereignty, is led by the government, and other groups collaborate to participate, collaboratively solve cyberspace-related problems and find development strategies within the framework of the United Nations W. Moritz, 2019 [5]. The latter is typical of Western countries, focusing on the needs of multiple stakeholders, weakening national sovereignty, and advocating for equality in participation. It is a bottom-up decision-making mechanism. The comparison of the two models is actually a contest between bottom-up and top-down decision-making mechanisms. In the practice of cyberspace governance, the United States practices multilateralism and has first-mover advantage, in fact the government has administrative jurisdiction to root servers, root zone files, the file system. Due to the international situation, the United States transferred its jurisdiction over IANA in 2016. Multilateralism was challenged by multilateralism, which is a kind of hierarchical order, the characteristics of network space more show the level of freedom and order, therefore, simple multilateralism or multilateral doctrine are not enough to achieve good governance in cyberspace.

2.3 Foreign experience in cyberspace governance

The United States is the initiator of the network technology and the superpower in the field of cyberspace, and has upgraded the cyberspace strategy to the national strategy. Its governance model is based on the multi-stakeholder model. On the one hand, it attaches importance to the development of technology-related enterprises and the protection of personal privacy data. On the other hand, the United States has established relatively complete legal documents, such as a network security review system. At the same time, at the content application level of cyberspace governance, the United States is constantly advancing the construction of military operations. This is a guarantee for national development that cannot be ignored or excessive. In addition, in the field of cyberspace, the United States has always been committed to technological innovation, from talent training to industrial investment, which deserves attention.

Singapore has a high degree of informatization, and the cyberspace governance has typical oriental characteristics, which can be used for reference in cyberspace governance of China. Singapore has established a sound organizational management system, policy system, and legal system in the cyberspace governance system. In 2015, the Cyber Security Agency of Singapore (CSA) was established to coordinate cyberspace governance and security matters. Its policy documents involved the construction of a secure cyberspace, the development of a dynamic cyberspace ecosystem, and the construction of a resilient key Information infrastructure, strengthening international cooperation in cyberspace and many other aspects. In 2017, the new cybersecurity bill was revised, and laws including the *Domestic Security Act, the Personal Information Protection Act, and the Abuse of Computer and Cybersecurity* Act were established to regulate cyberspace order. Singapore's cyberspace governance process has its own characteristics. The government's functional role is no longer "all-round", but "cooperative" Wei Wang, 2018 [6]. First, cooperation with ASEAN and other countries actively promotes the healthy development of global cyberspace. Second, tries to cooperate with domestic and foreign high-tech enterprises or colleges to absorb superior technology and professional personnel training methods. Third, Singapore cooperate with domestic people and explore national participation in cyberspace security governance.

The characteristics of cyberspace governance in Japan in the new era can be elaborated from three aspects. First, from the perspective of economic governance, Japan attaches great importance to the comprehensive upgrade of the application level of cyberspace, and the application and development of digital technology are laid out in an orderly manner under the escort of a series of policies. IT Comprehensive Strategy Headquarters promulgated the *Digital Procedure Law* in May 2019, and formulated the *New IT Policy Outline for the Digital Age* in June 2019. Japan accepted applications for the "information banking" industry in 2018, the first authentication enterprise was born in March 2019, and it is at the forefront of the business model that uses personal information to create value worldwide Man Ding, 2020 [7]. Secondly, from the perspective of security governance, marked by the formulation of the *Network Security Strategy* in 2013, Japan officially entered the era of ultra-high development of network security, with increasingly improved institutional policy documents and authority departments, and explored a network space governance model combining the use of public power and corporate self-discipline. Finally, from the perspective of the integration of economic governance and security governance, Japan's revised *Network Security Strategy* in 2018 first proposed the strategy of building a "cyber security ecosystem" in 2020, which is a mechanism for Japan's international policy coordination mechanism to force itself to proceed useful exploration of practice, This strategy not only accelerates the practice of cyberspace governance in Japan, but also conforms to the governance philosophy of international cyberspace.

On the whole, taking the above three countries as examples, we can see that developed countries have the following experiences to learn from in terms of cyberspace governance: first, establish a sound institutional guarantee mechanism, including sound organization and clear management organization positioning, complete laws and regulations, and sound policy documents; The second is to pay attention to the development and application of the application layer in cyberspace to construct the ecosystem of cyberspace. Third, upgrade the cyberspace strategy to a national strategy; Fourth, the government should work with enterprises and citizens to advance the project of cyberspace governance.

3 Challenges of our country's cyberspace governance in the new era

Cyberspace governance of China is facing both international and domestic challenges, which are embodied in the international environment, the scale of netizens, and cyber anomie.

3.1 The international cyberspace environment is complex

Between 2000 to 2009, the top Internet user growth not appears in developed countries, but in Africa, Latin America and other regions. It shows that more and more actors enter into the practice of the network space. The penetration rate of cyberspace in the fields of economy, security, individual power, and culture is getting higher and higher, and the cyberspace strategy has been upgraded to a national strategy. Various countries have coordinated and discussed the sovereignty involved in the three-tier structure of cyberspace governance structure. The level of technological development in the digital age is different Haimin Li, 2019 [8]. My country's international influence has become one of the key factors affecting the build of a community with a shared future in cyberspace.

3.2 The number of network users is surging and the structure is diversified

The 44th China Internet Network Development Statistical Report by China Internet Network Information Center (CNNIC) shows that the scale and penetration rate of China's Internet users have risen significantly from 2016 to 2019. By June 2019, the scale of China's Internet users has reached 854 million, with an Internet penetration rate of 61.2%, as shown in the figure 1. The distribution of network users in terms of age, occupation and gender shows different characteristics, and the network access methods are also more diversified. The large number of Internet users and the diversified structure pose a challenge to China's cyberspace governance.

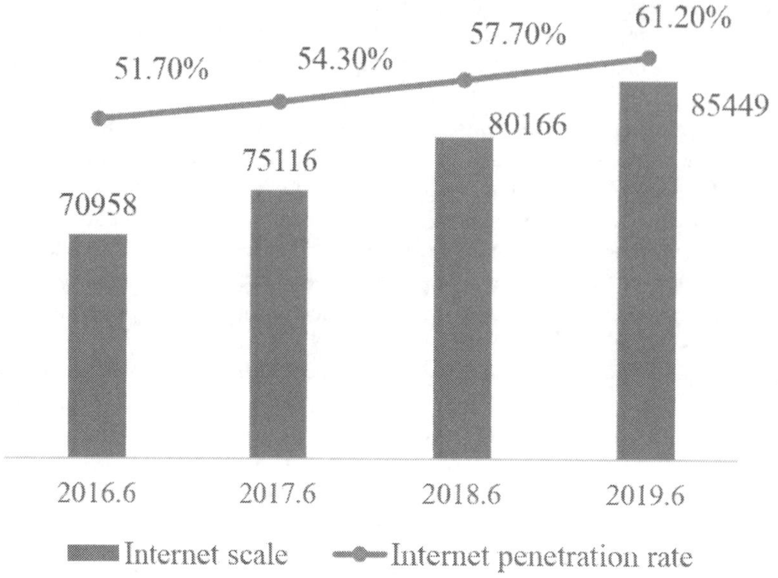

Fig. 1. The internet scale and penetration rate in China.

3.3 The number of cyber anomie climbs and the scope expands

Cyber anomie is mainly manifested in the form of cyber fraud, cyberattack, and cyber data leakage, and we are faced with both international and domestic cyber anomie. According to Reporting Center of Cyberspace Administration of China, the total number of acceptances of cyber violations and bad information in our country was 12.506 million in May 2019, while it rose to 15.199 million in May 2020. Although the relevant departments have taken certain measures in cyberspace governance, however the effectiveness is not optimistic. There is room for improvement in the role positioning and organizational management structure setting of departments and institutions, and the corresponding laws and regulations concerning cyberspace governance need to be further improved to provide guarantee for cyberspace governance.

4 Countermeasures of cyberspace governance in our country

According to the challenges we face in cyberspace governance, our country should take corresponding measures to deal with it.

4.1 Strengthen international cooperation and pay attention to internet technology innovation

Internet technology is the foundation of cyberspace and the driving force of national development. According to the “Global Connectivity Index” released by Huawei, information and communication technology (ICT) infrastructure investment is highly correlated with economic growth, and every 20% increase in ICT investment will pry 1% GDP growth. Therefore, in the exploration of cyberspace governance mechanisms, the first thing to consider is to improve the ability of Internet technology innovation, which is a consistent practice in the United States. First of all, we can refer to the practice of Singapore, draw on the technical strength and professional talent training experience of developed countries in national cooperation, and then we hope find technological breakthrough points, improve the strength of professional talents in China, build technological pioneers by top-notch innovative talents. Second, in terms of economic development, application and security management of cyberspace, Japan is at the forefront of in the word. We should also strengthen cooperation with other countries, carry out multinational cooperation on the basis of trust, promote the exchange of experience and consultation on international rules, and jointly build a community of shared future in cyberspace.

4.2 Innovate the governance model of cyberspace and coordinate with multiple actors to participate

In the process of cyberspace governance, we should exert public power, at the same time give full play to the initiative of other actors. Exercising public power means that cyberspace governance proceeded by a powerful management organization. The function of the organization is not only to supervise, but also to take the initiative. It is also a key force for coordinating the participation of enterprises, democracy, civil society and other actors. To exert the power of other actors, we must adhere to the principle of industry self-discipline. The self-discipline of the Internet industry is the weak link of our cyberspace governance, China should give them rights and space to give play to their initiative and creativity, and provide them guidance in key links Tianshu Que,2018[9].

4.3 Formulate policies and laws to ensure system security

Reducing cyber anomie and promoting the development of the content application layer in cyberspace are inseparable from a sound institutional system. The United States, Singapore, and Japan have rich experience in this area and are worth learning. Therefore, on the one hand, we should learn from foreign experience, explore a policy system suitable for China's national conditions, and form a comprehensive system covering infrastructure construction, cyber security space ecosystem, and foreign cooperation. On the other hand, we should improve relevant laws to guarantee the exercise of rights and obligations of all actors.

5 Summary

With the development of Internet technology and the expansion of its application scope, my country's cyberspace governance is facing a more complex international environment, a larger scale of netizens, and more prominent cyber anomie. Cyberspace governance is facing unprecedented challenges. First, China should continuously strengthen international cooperation and focus on cultivating our own technological advantages. Second, we also need to improve the organization and management system of cyberspace, take the government as the lead, cooperate with enterprises and the public to jointly govern, and promote industry self-discipline. Finally, China must establish a perfect policies and legislation to provide institutional guarantees for the practice of cyberspace. Thus, China should continue to promote the integration of cyberspace and physical space in the new era, and create a favorable situation in cyberspace.

References

1. C Y Lu, *Cyberspace Governance: Power Games, Conceptual Evolution, and China's Strategy*, J. Global Review. **1**, 117-134(2016)
2. Z Z Xu, *The Task and Challenge of Cyberspace Governance*, J. Chinese Cadres Tribune. **1**, 36-37(2020)
3. H G Zhou, *On Strategic Stability in Cyberspace*, J. Global Review. **1**, 21-34(2019)
4. B G Cui, Jinhe Liu, *On the Platform Governance in Cyberspace*, J. Global Media Journal. **7**, 86-101(2020)
5. W. Moritz, J. Vytautas, *Securing cyberspace: How states design governance arrangements*, J. Governance. **2**, 259-275 (2019)
6. W Wang, Singapore*'s Cyberspace Governance and Its Enlightenment to China*, J. Pacific Journal. **26**, 35-45(2018)
7. M Ding, *Digital Economy and Cyberspace Governance Strategy of Japan*, J. Contemporary Economy of Japan. **1**, 01-12(2020)
8. H M Li, *The Digital Silk Road and the Reconstruction of Global Cyberspace Governance*, J. International Forum. **21**, 15-29(2019)
9. T S Que, *Government Regulation and Good Governance in cyberspace: Logic, Mechanism and Path Choice*, J. Contemporary World and Socialism. **4**, 184-192(2018)

On the Main Problems and Countermeasures of Green Travel in Wuhan

Liang-Bin Cheng* *and Mengke* Hu

School of Marxism, Huazhong University of Science and Technology, Wuhan 430074, China

Abstract. At present, the main problems of green travel in Wuhan are as follows: serious segmentation, insufficient promotion of green new energy vehicles, single way of QR code scanning, low public participation, etc. In order to solve these problems and further improve the green travel rate and air quality in Wuhan, the following countermeasures need to be adopted: improving the top-level design, strengthening tax relief and financial support, establishing a market-oriented incentive mechanism, developing green public transport, regularly making statistics and releasing relevant data of green travel, strengthening the education and publicity of green travel, etc.
Keywords: Green travel, Problems, Countermeasures, Wuhan.

1 Introduction

In 2013, China suffered from severe fog and haze, with the average number of haze days reaching 29.9 days across the country, spreading to 25 provinces [1]. In that year, Wuhan had recorded low number of days with good air quality, with a rate of only 44.1 percent. This situation has a great impact on people's health, work and life. Obviously, the cause of haze weather is closely related to people's production and life style (especially their travel mode). In recent years, Wuhan has witnessed a rapid growth of motor vehicles. By the end of 2019, the number of motor vehicles in Wuhan has reached 3.3 million, an increase of 14.6% over 2017. This has led to an increase in pollutants such as carbon monoxide, nitrogen oxides, PM2.5 and PM10. According to calculations, the current contribution rate of motor vehicles in Wuhan to the PM2.5 concentration in the atmospheric environment has reached 27%[2]. Therefore, it has become the common voice and basic appeal of the people to vigorously promote green travel, greatly improve air quality and effectively curb the use of non-green new energy vehicles.

2 The main problems of green travel in Wuhan and its causes

2.1 The fragmentation is serious and the operation mechanism is not smooth

In May 2019, the Ministry of Transport, the Central Propaganda Department and other 12 departments and units issued the "Green Travel Action Plan (2019-2022)". This means that

* Corresponding author: chengliangbin@hust.edu.cn

these 12 departments and units are the governance principal parts of green travel work. However, the action plan did not mention anything about how the 12 departments and units will communicate and coordinate their responsibilities.

In Wuhan, green travel mainly refers to travel within the city, which may not involve so many departments and units, but at least involve 9 departments and units, such as Transportation Bureau, Publicity Department, Development and Reform Commission, Environmental Protection Bureau, Public Security Bureau, Urban Management Bureau, Land and Resources Bureau, Public Transport Group, Metro Group, etc. But so far, Wuhan has not set up relevant organizations and coordination agencies, which make it difficult to fully integrate the advantageous resources of these departments and units, and even more or less appears the situation of separate governance [3]. For example, WeChat mini program of "Low-carbon Military Transportation" launched by the National Development and Reform Commission is well-targeted, effective, time-sensitive, and affordable. It does not fall behind other similar programs, however, it is not well known in Wuhan and the number of registered people is small, which obviously has a great relationship with insufficient publicity.

2.2 Insufficient promotion of green new energy vehicles

By the end of 2018, the number of new energy vehicles in Shanghai had reached 240,000, and the total number of registered new energy vehicles in Shenzhen had reached 270,000. In April 2019, Tianjin also had 125,000 new energy vehicles. Relatively speaking, the growth of Wuhan's new energy vehicles is relatively slow. As of February 2019, there are only 53790 new energy vehicles in the city [4].

In terms of green new energy buses, Hangzhou took the lead in introducing new environment-friendly buses such as gas-electric hybrid and natural gas as early as 2008. By the end of 2016, the proportion of green buses in Hangzhou had reached 96.98%, leading the country. In 2017, Shenzhen became the first city in the world to make its buses 100% electric. The city of Wuhan did not start purchasing pure electric buses until 2014. In 2015, it stopped purchasing diesel buses and started to replace old vehicles with pure electric or natural gas buses. By the end of 2018, the proportion of Wuhan's green new energy buses (including pure electric buses and natural gas) have reached 62%, and it is expected to reach 80% by 2020 [5]. Compared with Shenzhen, Hangzhou and other cities, there is still a big gap in these data.

In addition, in 2018, gasoline vehicles accounted for 91.67% of Wuhan's motor vehicles, diesel vehicles accounted for 7.70%, and the two together accounted for 99.5% [6]. This data reflects that the promotion of green new energy vehicles in Wuhan has just started, and there is still much space for development in the future.

The main reasons for the insufficient promotion of green new energy vehicles in Wuhan are as follows: First, insufficient financial subsidies. At present, when Wuhan enterprises purchase a new energy vehicle, the state subsidy and local government subsidies together account for about 60% of the car purchase, but the proportion in Xiangyang is about 80%. Second, the supporting facilities are not perfect. Due to urban development and transformation, some bus parking yards have been demolished, making it impossible to centralize charging and parking of new energy buses at night.

2.3 Public transport code scanning mode is relatively simple

At present, Shenzhen has realized seven modes of travel payment, including one-way tickets, national transportation all-in-one cards, UnionPay Quick Pass, etc. On the basis of traditional payment methods, Hangzhou has added a variety of payment methods such as subway official APP, Alipay APP scanning code and so on. Beijing also launched the "One Code" in public transportation on May 16, 2020.The "One Code" service breaks the barriers between different subway and bus scenes, and realizes the interconnection among various public transportation apps (Beijing Bus APP, Yitongxing APP, Beijing One Card APP), that is, passengers can use any APP code to take the bus or subway, saving you the trouble of downloading different applications.

Relatively speaking, mobile payment in Wuhan is a bit inferior. There are only two kinds of mobile bus payment, one is Alipay scan code bus, the other is Wuhan pass and other physical card payment card; besides, the metro only supports payment with physical cards such as Wuhan Tong and exclusive apps. Although Wuhan Metro launched the mobile phone scan code to purchase tickets and scan code gate functions in December 2017, it only supports mobile phones with NFC function. Passengers also need to download and install the "Metro New Era" APP before they can use it after registration and binding. In other words, the Wuhan Metro does not support UnionPay Quick Pass, Tencent ride code, Alipay scan code and other ways of crossing the gate. In addition, as of June 2019, a total of 245 cities in China have been connected by the all-in-one transport card. At present, buses and ferries in most parts of Wuhan have joined the system and enjoy a 20% discount, but Wuhan Metro has not yet joined the system.

2.4 Insufficient publicity of green travel and low public participation

Since China joined the 922 Green Travel campaign, various activities have been carried out throughout the country to mobilize and enhance the public's awareness of green travel. From September 2013 to September 2019, the "Green travel, smooth Beijing" traffic propaganda team of Beijing Municipal Commission of Transport held more than 1,700 activities for people from all walks of life, and the number of participants reached 300,000. Compared with Beijing, Wuhan's green travel campaigns have some problems in terms of mobilizing the public to participate, such as a small range of audiences and a low level of public participation. At present, there are only more than 300 volunteers in the green travel publicity campaign, and the publicity scope and intensity are limited. As a result, the publicity target group is only a small number of citizens, which has not been widely spread to all groups in the whole society[7]. In particular, it has not played a propagandistic and educational role on the main force of green travel such as salaried employees.

3 Countermeasures for green travel in Wuhan

3.1 To improve top-level design and strengthen system governance

We can learn from the Hangzhou "Five Waters Co-governance Office" with relevant experience in institutional setting, the Wuhan Green Travel Work Leading Group should be established under the Wuhan Environmental Protection Commission to coordinate and manage the city's green travel related work. It is recommended that the deputy mayor in charge of transportation or the environment concurrently serve as the leader of the green travel work leading group. In terms of constituent units, the Transportation Bureau, the Environmental Protection Bureau, the Urban Management Bureau, the Public Security Bureau, the Development and Reform Commission, the Land and Resources Bureau, the Publicity Department and other departments with green travel development planning function, publicity function or law enforcement power shall be included, and the relevant persons in charge shall be the team members. An organizational structure with centralized functions and unified responsibilities shall be formed, whose functions are to guide, coordinate, supervise, assess and publicize, and carry out comprehensive management for those that no one has managed before, dare not manage, do not want to manage or cannot manage, and make overall arrangements for green travel work, which shall be implemented by relevant departments. It is recommended to include a green travel project in the "Wuhan Embracing Blue Sky Action Plan 2020", so that green travel work should be included in the annual specific plan (the plan did not explicitly mention green travel in previous years).

In the "Wuhan Motor Vehicle Exhaust Pollution Prevention Regulations" (approved by the Standing Committee of the Municipal People's Congress on September 17, 2014, and formally implemented on March 1, 2015), clear regulations on green travel are given: "organizations , groups, enterprises, institutions and other organizations shall strengthen motor vehicle exhaust pollution control laws, regulations, and civilized traffic, green travel propaganda", "The Municipal People's Government should create conditions to facilitate citizens' green travel." As local regulations, these regulations are relatively advanced nationwide, and their ideas are relatively advanced, reflecting the Wuhan Municipal Party Committee and government still attaches great importance to green travel. However, in order to make these regulations come true, it is necessary for the municipal Party committee and the municipal government to carry out strong system governance.

3.2 To strengthen tax reduction and financial support

At present, with the continuous increase in operating costs such as equipment renewal and maintenance, public transportation companies generally suffer from a shortage of funds. Taking the subway as an example, according to the statistics of the "China Urban Rail Transit 2018 Annual Statistics and Analysis Report", the average operating revenue and expenditure ratio of urban rail transit in 2018 was 78%, and only four cities including Hangzhou, Qingdao, Shenzhen, and Beijing had operating income more than 100%, at least not at a loss. The rail transit operations of other 24 cities, including Wuhan, have a certain degree of loss, and they need to rely on financial subsidies to maintain their operations. Therefore, Wuhan can appropriately increase financial support for public transportation in accordance with the development of the public transportation sector under the conditions of local financial strength.

In addition, it is recommended to set up a special fund for green travel development, increase financial subsidies for public infrastructure construction, and realize 100% new energy for buses[8]; increase funding guarantees for slow-moving system improvements, and encourage and support shared bicycle companies to enter; increase financial support for technology industries such as new energy and information, and encourage relevant companies to develop green and pollution-free vehicle fuels through financial subsidies and tax reductions, so as to produce more energy-efficient and low-carbon environmentally friendly vehicles; increase taxes and fees on the traditional energy-intensive automobile industry, while increasing subsidies for the purchase of new energy vehicles to guide the concept of green and sustainable consumption.

3.3 To strengthen government-enterprise cooperation and establish market-based incentive mechanisms

First of all, Wuhan can learn from Japan's "Environmental Protection Points System" (2009), Guangdong's "Carbon Inclusive" (2015), Nanjing's "Riding Points" (2016) and Shenzhen Yantian District's Carbon Coin System (2016). The specific approach is to further improve the "Low Carbon Military Games" platform launched by the Wuhan Development and Reform Commission in June 2019 to improve the platform's positioning, convenience and timeliness. One is to further enhance the positioning of "Low Carbon Military Games", not only for the carbon neutrality of the military games, but also to include the construction of an "ecological Wuhan" after the military games; the second is to improve the convenience and timeliness of the platform. For example, the platform's activities such as "organic light food, favorite accompanying cups" and other activities need to be photographed and uploaded, accounting for 40% of the total activities, and these activities need to be manually audited before they take effect. The review process will not only incur certain manpower and material costs, but also affect the timeliness of points exchange. Regarding the two activities of "low-carbon hotels" and "purchasing green home appliances", we can learn from the experience of Yantian district in Shenzhen. The government confirms green hotels and green home appliances sales brands through inspections and establishes contact with them. If the public goes to the corresponding location to generate transactions, then you can directly scan the QR code to get points.

Secondly, in view of the key and prominent problems in Wuhan's green travel, corresponding applications will be launched in a timely manner, by mobilizing the enthusiasm of the broad masses of people, internalizing the externalities of green travel, and solving problems in a market way. For example, in response to the important impact of private cars on green travel rate, air quality, and traffic congestion, the Shenzhen Traffic Police launched "Green Travel" (entered on WeChat and Alipay platforms in September 2017) to encourage private cars to travel green (that is, no or less use, disabled during peak hours, etc.). As the "second headquarters" of many technology companies such as Alibaba and Tencent, Wuhan can rely on existing platforms such as Alipay, Tmall and WeChat to effectively graft Internet technology and green travel, and build various platforms for green travel in Wuhan based on market-oriented means.

3.4 To develop green public transport

The new energy buses are basically pure electric buses, with zero carbon emissions and only 65% of the driving cost of diesel buses. They are not only economical and environmentally friendly, but also have low odor and noise, and have extremely high cost-effective. Take Shenzhen as an example, in the past, buses and taxis only accounted for 1% of the total number of motor vehicles, but the pollution generated accounted for more than 30% of the city's motor vehicles. After continuing to promote new energy vehicles, "take Shenzhen Bus Group as an example, in the past year, pure electric buses saved about 116,000 tons of fuel and reduced carbon dioxide emissions by about 360,000 tons." [9]

It can be seen that although the increase in the ratio of green and new energy public transportation has no effect on the improvement of green travel rate (according to the current calculation method), but it has a great impact on air quality. Therefore, the city of Wuhan should fully realize the importance of developing green buses, and learn from the earlier and faster development of new energy buses in cities such as Shenzhen and Hangzhou, strengthen the application of new energy technology research and development, accelerate the development of diesel buses, and strive to achieve the goal of 100% of the green and new energy public transport ratio as soon as possible.

3.5 To make statistics and release relevant data of green travel regularly

At present, the data system about green travel in Wuhan is incomplete, and it can't be updated in time. For example, the core data of green travel "green travel rate" so far, the author has not found the official data. However, in 2017, the Ministry of Transport clearly stated in the development goals of the "Opinions on Comprehensively and Deeply Promoting the Development of Green Transportation" that "by 2020, the proportion of green travel in the central urban areas of large and medium-sized cities should reach over 70%". In 2018, the Wuhan Municipal People's Government clearly pointed out in "Wuhan Urban Construction

Green Development Implementation Plan (2018-2020)" that it will take three years to "increase the urban public transportation, bicycle, walking and other green travel sharing rate by more than 10%." According to analysis, the "green travel ratio", "green travel sharing rate" and "green travel rate" actually mean the same thing here.

If we don't know the green travel rate of Wuhan at that time, how can we make a targeted plan to ensure that the goal of reaching 70% or more in 2020 is achieved (if it has not been reached yet)? If we don't know the green travel rate in 2018, it is not rigorous or clear to say that the green travel rate will be increased by more than 10% by 2020. Because the base is different, the 10% increase rate or percentage point is different (such as 50% of 10% is 5 percentage points, and 10% of 90% is 9 percentage points).

Currently, the governments of Beijing, Shanghai, Chongqing, Nanning, and Bao'an District of Shenzhen all regularly release relevant data such as green travel rates (see Fig.1). These methods have successful experiences of Wuhan to go by.

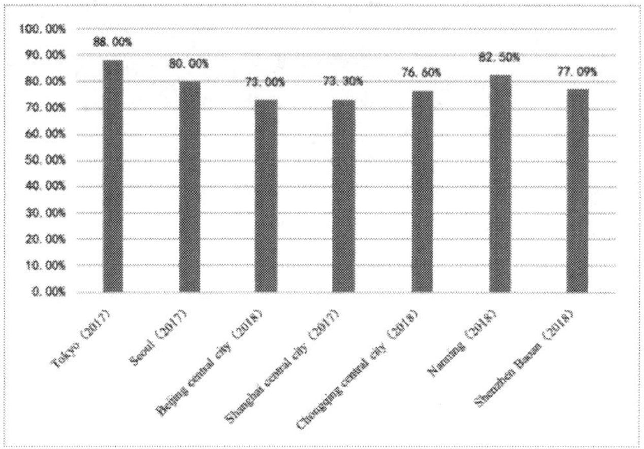

Fig.1. Distribution of green travel rate in some cities (or districts) at home and abroad. (Source: These data are collated from network resources.)

3.6 To strengthen the education and publicity of green travel

First of all, we must play the basic role of school education. The Municipal Education Bureau takes the lead in bringing green travel into the education of universities, middle schools and primary schools, so that the newly grown generation will become the pioneers of green travel and active practitioners of the "two mountains" theory. Different education forms and different education plans can be adopted for students of different levels, different ages, and different genders; the first classroom and the second classroom can be organically combined.

Second, we must give play to the leading role of news media. Relying on the green travel promotion month and bus travel Publicity Week held in September every year, Wuhan can put up propaganda posters and broadcast short films in public places such as underground passages and bus station billboards, and carry out activities such as selecting green travel pioneers and collecting typical cases by using traditional media and network channels. At the same time, it is suggested to regularly broadcast the public service advertisement of green travel in the mainstream media, and timely launch the application of promoting green travel, so as to normalize and concretize the publicity work of green travel.

Finally, we must play the key role of communities and non-governmental charity organizations. It is recommended that each community be equipped with volunteers of green travel promoters, and dedicated persons will spread the concept of green travel to different groups in the society through the distribution of brochures and other forms; learn from the successful experience of the "green travel, smooth Beijing" traffic propaganda group, and cooperate with "Wuhan green river city" non-governmental environmental protection organizations to form the Wuhan green travel propaganda group, which is composed of experienced volunteers and traffic workers who are struggling in the front line. We can popularize green travel knowledge, improve environmental awareness and form a unique ecological culture brand through vivid real cases and a wide variety of teaching forms [10].

References

1. Information on http://jl.cnr.cn/ltrt/201312/t20131230_514527688.shtml.
2. R. J. Chen, H. Y. Xie, Y. C. Lin. Journal of China University of Geosciences (Social Sciences Edition), **18**(04):95-110 (2018)
3. J. Xia, S. D. Qin, Hubei Agricultural Mechanization, (06),39-43(2013)
4. Information on http://m.evpartner.com/news/detail-46624.html
5. Information on https://www.sohu.com/a/290201525_760385.
6. Wuhan Ecological Environment Bureau. 2018 Wuhan Motor Vehicle Pollution Prevention and Control Annual Report [R] (internal data). Wuhan Ecological Environment Bureau. 2019: 4.
7. Y. Li, Social Sciences Front, (06),265-268(2017)
8. T. Q. Chen, Enterprise Reform and Management, (08),220-222(2019)
9. Information on https://www.sohu.com/a/290201525_760385.
10. R. R. Yang, R. Y. Long, Journal of Wuhan University (Philosophy and Social Sciences Edition), **67**(05),13-19(2014)

Study on the Right of Residents in Urban Residential District

Feng Guo*, *Haocheng* Sun, *Yuye* Chen *and Jiangning* Fang

School of Intellectual Property Rights, Nanjing University of Science and Technology, Nanjing, Jiangsu, China

Abstract. Due to the lack of the urban household registration or property owner, many residents of urban residential areas in China have been in the state of lack of rights. The policy of "Equal rights for home tenants and owners" cannot play a practical role due to the lack of legal norms. "Home ownership" and "Right of residents" contain different contents. That is, "equal access to basic public services and social welfare of the place of residence, as well as the right to participate in decision-making, management and supervision of public affairs in residential communities and residential areas due to the fact of stable residence." The rights of the occupants is different from housing right and habitation right. The right of the occupants has not only legal basis, but also practical basis.
Keywords: Right of residents, Right of housing, Right of habitation, Equal rights for home tenants and owners

1 Introduction

In China's cities, about 30% of the resident population are from other countries, who need citizenship. In a few megacities, half of the non-native migrants have lived there for more than five years, and about 20 percent have lived there for more than 10 years.[1] Such a large number of non-native migrants live in local cities by renting, and the tenants of residential buildings are in the dilemma of "home ownership" for a long time, but unable to fully integrate into the local civil society and enjoy the rights and interests of residents[2].On the one hand, according to the existing property management laws and regulations, as non-owner residents cannot participate in the owner governance activities of the residential community. On the other hand, according to the existing household registration policy, as unregistered foreign residents, they cannot equally enjoy the public welfare of the residence, nor can they participate in the grassroots democratic political activities of the local community. The fact that the occupant cannot change the separation of identity right and property right makes the public interest attached to the household registration and property right disappear because of the tenant's lease behavior.

In the report of the 19th National Congress of the Communist Party of China on strengthening the social security system, it was proposed to "accelerate the establishment of a housing system with multi-agent supply, multi-channel security and simultaneous rent and

* Corresponding author: fl-fl@263.net

purchase". Both rent and purchase simultaneously means to rent, and the tenant can best highlight the property that the house is used to live in. The implementation of the policy and the objective of the policy have become the watershed of the historical reform of housing system in China. However, there is no law to define the specific rights and interests of the same right of rent and purchase in China. The relevant laws and regulations aim to regulate the chaotic housing rental market and fail to focus on the series of rights changes caused by the lessee's renting. The legal essence of the same right of rent and purchase is to endow the lessee and the owner with the same rights in the field of basic public services [3], but the policy of the same right of rent and purchase cannot provide basic guarantee for the rights of residents. Under the constitutional spirit of respecting and protecting human rights, the rights and interests that residents should enjoy are always in a vague field.

2 The proposal of the concept of "right of habitant"

The inhabitant right refers to the right of residents to enjoy the basic social public service and social welfare equally, and to participate in the decision-making, management and supervision of the public affairs of the residential community and residential quarters due to the fact of stable residence. The right of habitant is different from the housing right and the habitation right. The existing theories and practices of the housing right and the habitation right are not applicable to the actual residents under a large number of residential lease contracts in China, that is, the residents are different from the subject of housing right and habitant right. The proposal of the concept of residence right is the scientific practice of the second generation of human rights of the United Nations in China's law. The content of the right of inhabitant is related to China's specific national conditions and social reality, which is more abundant and more difficult to define than the content of housing right.

As far as occupant rights are concerned, which is similar in nature to the right of housing in international human rights conventions but different in contents. The right of housing is the right of minimum living security. The right design is based on the protection of human rights and pays attention to the survival and development of natural persons. However, the right of housing which is subject to economic factors will be guaranteed differently according to the national conditions of different countries. So far, the right of housing has not been recognized and protected in China's law. In China's social environment, even if the right of housing is recognized, the right also needs to be converted in Chinese way in order to be compatible with the household registration system and housing system.

The right of residence is one of the usufructuary rights in other property, which is used to adjust the housing usufruct relationship between equal civil subjects. From the perspective of the protection of real right, the right of habitant is based on the protection of the basic right of residence of the right holder, and it does not concern the right holder's participation in other aspects of citizen's life. Accordingly, the protection of the right of habitant focuses on some aspects, which cannot cover all the rights and interests of the oblige. Therefore, other rights protection systems and theirjoint functions are needed to guarantee the citizens' enjoyment of grassroots democratic rights and social welfare rights.

Thus, the right of housing and the right of habitation cannot abstract the right of habitation and form the complete content of the right of habitation. The concept of habitant right is put forward due to the necessity of extension of the above theory and the urgency of habitant right protection.

The right of habitant means the right enjoyed by the habitant. From the normative analysis of the right, the subject of the habitant's right is the actual habitant in the migrant population, without the local household registration, but it has a relatively stable living fact. By paying rent or property fee, they have the public interests and services attached to the buyers or property owners. In fact, the residents get a free physical and geographical space, the residence and activities are not temporary or temporary for special purposes, but relatively long-term living, working and living in the local area consumption, entertainment and labor.

The object of the habitant's right, that is, the object of the inhabitant's right, is not limited to the rented house, not limited to the autonomy of the grassroots mass organization, but also includes the inheritance of the public interest enjoyed by the property owner because of the stable living fact. The legalization of the concept of habitant rights not only enables the majority of residents to realize the basic housing goal of "home ownership", but also enables them to live and work in peace and contentment [4]. The content of the right of habitant includes many aspects. At present, the trend of "renting and purchasing simultaneously" in China's housing reform policy can reflect part of the contents of the residents' rights, and the slogan of "renting and purchasing the same rights" can better reflect the value orientation of the residents' rights. The core of habitant rights is the non-discrimination of identity, the indifference of rights and the equality of status. The right of inhabitant is similar to the second generation of human rights "housing right" proposed by the United Nations. However, due to the household registration system and housing system in China, the right of habitant has its own special features. With the development of the social system, the content of the right of residence will change. In terms of China's current national conditions, the right of residence specifically covers the following aspects: children's education, medical security, employment support, pension services, social welfare, and political participation [5].

The right of the inhabitant is relative to the place of residence. When a natural person changes his residence, it involves the elimination of the right of inhabitant. The acquisition of the right of inhabitant relative to the new residence is conditional, that is, in line with the principle of "the right to go with the place of residence". When a natural person migrates the original habitat to a new place to live, he will lose his original inhabitants, not shall enjoy the right of residents both at the same time, residents can only obtain the actual residence of the public service and social welfare, to prevent improper occupy and waste of public resources.

3 Legal principle of the right of habitant

The right of habitant is separated from the right of housing and the right of habitation, which not only has the unique right connotation, but also has the practicallegal principle.

First of all, the right of habitant accords with the legitimacy of human rights protection. One of the important reasons for the legitimacy of the establishment of the system of residents' rights is that it has further improved the protection system of people's right to subsistence and development. From the perspective of human rights protection, the function of habitant right system is to protect people's right to survival and development. The system of habitant rights contains the protection of the subject's right to subsistence and development, which is in line with the principle of minimum living guarantee of housing right in the United Nations Convention on human rights and the interests of the usufructuary right holders in the property law. The right of habitant can well reflect the uniqueness of the right of development, in the implementation by identity, eliminating property gives residents shall be equal opportunities for development. The prerequisite for the existence of the right of habitation is the guarantee of freedom of residence, the most representative of which is the provision of article 22 of the Japanese Constitution: "[freedom of residence, migration and choice of occupation, freedom of moving to a foreign country and seceding from nationality] a. Everyone has the freedom to live, to move and to choose his occupation, so far as it is not contrary to public welfare. b. The freedom of any person to emigrate or to renounce his nationality shall not be infringed." The establishment of the system of residents' rights can make the residents better survive in their residence, and obtain the development conditions which cannot be obtained by the household registration and housing system before.

Secondly, the right of residence conforms to the ultimate goal of the development of social welfare system. Different from the minimum living guarantee nature of housing right and the real right interest to make the best use of everything pursued by the right of residence, the right of habitant has jumped out of the right level of the minimum living guarantee and has a higher level of social right demand on the basis of the exercise of usufruct. The subject of the right of habitant does not choose to live in a city because of poverty, but chooses to live in a development place in pursuit of a better life goal, which is consistent with the goal of beauty, happiness, equality and justice pursued by social welfare policies. The emphasis on Residents' rights means opposing social exclusion and promoting social integration. It requires the government to use public power to intervene in the unfair distribution of public resources and eliminate the differences in interests of residents in obtaining basic public services provided by the government due to their status.

Finally, the right of habitant embodies the objectivity of the legal value. This can be seen from three aspects. Firstly, the system of the right of residents has objective justice value. Since ancient times, the justice value of law is often regarded as not only the highest value of law, but also the highest value of human society. The system of habitant rights creates new right distribution way, namely by giving corresponding conditions or any natural person having a corresponding duty in a particular region of the identity of the residents, and bases on this identity, they can enjoy various public services and rights that are convenient for them to continue to live and contribute to the development of the region, so as to promote local development. Secondly, the occupant right system has the objective value of freedom. The ultimate goal of the right to create residence is that the natural person who has acquired the residence status can enjoy the corresponding freedom. The system of habitant right is based on the premise and purpose of freedom, and it is also the way to realize freedom. Thirdly, the occupant right system has objective and equal value. Equality, like freedom, is the core value of modern society. Right of habitant breaks through the shackles of property right system and household registration system, which enables the migrant population in the same area to enjoy the same civil rights of local residents, and solves the problem of residents' right demands caused by the separation of status right and real right for a long time.

4 Basis for the realization of habitant rights

The history of the development of human rights provides a practical basis for the construction and operation of the legal system of residents' rights. As mentioned above, the right of resident is similar to the right of housing in international human rights covenants. The concept of housing right originates from the expression of "the right to adequate housing" in international human rights covenants, as shown in article 11, paragraph 1, of the International Covenant on Economic, Social and Cultural. Rights: "the states parties to the present Covenant recognize the right of everyone to an adequate standard of living for himself and his family, including adequate food, clothing and housing and to the continuous improvement of living condition." On December 12, 1991, the United Nations Committee on economic, social and cultural rights held that the right of housing is an important right in social rights. In order to solve the problem of people's vague understanding of the right of housing, it committed itself to providing a basis for a unified understanding, and adopted General Comment No.4(1991) The Right to Adequate Housing, namely The General Comment No. 4: The Right to Adequate Housing. The right of housing refers to the right to be housed and to live in dignity. The most critical thing is that the right of housing is the right of the Government, which has the obligation to provide citizens with suitable housing and to improve their living conditions. Under this theory of human rights, the content of welfare doctrine contained in the right of inhabitant, like economic and social rights, depends on the state's action, that is, the realization of rights is the obligation of the state.

Under the framework of international human rights and protection under the framework convention, the right of housing cannot be ignored as a basic human rights, such as Spain, Portugal, Switzerland, Finland, Russia, South Africa and other countries have confirmed in the constitution, making special housing or legal right to affordable housing, such as the United States, Germany, Australia, Japan, Singapore and other countries. Although China's constitution does not explicitly mention the right of housing, the constitutional principle of respecting and protecting human rights and the fact that China is a state party to the International Covenant on Economic, Social and Cultural Rights makes it possible for the right of housing to be incorporated into the Constitution and for laws on housing security to be enacted, which is also the need for China to fulfill its international obligations. The legislative practice of the above countries provides experience for the construction of the legal system of the right of habitant. The right of inhabitant, which is related to the right of housing, is also facing the opportunity of legalization. Some scholars believe that "the basic right status of housing right should be confirmed at the constitutional level, so as to form an objective legal value order that standardizes the legislation, administration and judicature."[2]

However, at present, the judicial organs cannot directly apply the basic rights of the Constitution in our country. That is to say, the court cannot accept constitutional rights litigation [6]. However, the newly added right of habitation in the civil code of China provides the legal and material basis for the habitant's right. The right of residence promotes the convergence and development of traditional property rights and modern residential rights, further improves China's property law, more conducive to meet the people's basic living security, provides a legal system guarantee for solving housing problems with multi-level property rights structure [7], and provides a legal basis for the protection of the rights of habitant.

In practice, the right of habitant system is mainly applicable to the economic assistance of divorce, the needs of pension security, and the housing problems of minors. However, after improvement, the purpose of establishment of the right of residence has also been extended from basic living needs such as support and maintenance to housing investment, which has also played a positive effect on commercial activities. The residential right system is a breakthrough and extension of the current legal norms. The expansion of its scope of application is in line with the objective needs of social development. Undoubtedly, the right of habitation has become an important way to realize the right of habitant, which provides a realistic material basis for the protection of the right of habitation. There are conflicts between the right of habitant and the right of habitation. The starting point of the right of inhabitant is the interest of human development, which is different from that of the right of abode.

The improvement of grassroots social governance system and governance capacity can provide a good environment for the realization of habitant' rights. The residents enjoy the right of inhabitant because of their actual residence, which not only inherits the democratic rights of the owners in the community, but also obtains the public resources and social welfare enjoyed by the residents of the residence. However, the realization and guarantee of all this cannot be achieved without the high recognition and active implementation of grassroots autonomous organizations. The Fourth Plenary Session of the 19th Central Committee of the Communist Party of China proposed "to build a new pattern of grassroots social governance and improve the institutionalized channels for the masses to participate in grassroots social governance." It can be seen that the current society has changed from the "management" mode at the grass-roots level to the "governance" new situation. The community assumes the responsibility of social security and public services at the bottom level. As the grass-roots organization for the implementation of the government's livelihood policies, residents' committee should become a powerful organization for the protection of residents' rights. "For the society, the autonomy of the business council is the first priority, which is endowed by the society with high expectations of safeguarding the rights and interests of owners, promoting grassroots democracy and building civil society, thus gaining high social legitimacy."[8] As an autonomous organization of the owners of urban residential communities, the owners' committee can directly reflect the public opinion and convey the valuable opinions of the residents of the communities on the protection of the residents' rights. Through the cooperation of the residents' committee and the industry committee, the residents can get rid of the shackles of household registration and owners' identity, and truly realize the residents' rights.

5 Conclusion

Since World War II, the practice of the second generation of human rights "right of housing" of the United Nations, and the implementation of the current "renting and purchasing simultaneously" policy in China, have made the legal sense of the concept of occupant rights precipitate. Through the definition, analysis and deconstruction of the concept of "right of habitant", it is proved theoretically that the right of habitant is independent of the right of civil residence. The legitimacy and rationality of the construction of rights of habitant can not only solve the shortage of housing policy, make the same right of renting and purchasing feasible, but also effectively promote the equalization of basic public services, boost the legalization, and solve the new contradictions aroused by social injustice in the new era.

Acknowledgements

This research was financially supported by: the National Natural Science Foundation of China-- The Rule-following Behavior of Owners and its Impact on the Governance of Urban Residential District in China (71774082) and JiangSu Undergraduate Training Program for Innovation and Entrepreneurship (201910288159Y).

References

1. Y. Qu, G. Zhang. What are the top priorities in the 14th five years plan? Experts outline the key points for you [EB/OL]. People's Daily ,2020-08-28[2020-09-04]. http://www.people.com.cn/BIG5/n1/2020/0828/c32306-31841145.html.
2. J. Liu, Y. Xing. Study on the Realization Mechanism of Public Service Equalization between Buyer and Tenant, SER, **35** (05), 51-59, (2018).
3. H. Xie. Legal Implications of the same Right of Lease and Sale and its Realization Approaches, **26** (27), 100-102, (2017).
4. Y. Zhang. Policy objectives and legal structure of China's Housing Leasing Law under the same right of renting and purchasing, **63** (16), 86-89, (2019).
5. J. Chen, Y. Wu, The Possible Conflict between the Housing Rights and the Rights to Public Services Availability in the Process of "Equal Rights between Buyer and Tenant"— Renting for Living or Renting for Rights, **63** (2), 44-56, (2019).
6. H. Huang, Y. Pan. Discussion on housing right and its legislative guarantee, **35** (01), 82-84, (2013).
7. H. Quan, JJU. Discussion on the system of habitation in The Real Right of Civil Code, **39** (02), 119-121, (2020).
8. L. Xiao. The Asymmetric Legitimacy: A Comparison between Resident's Committee and Homeowners' Committee, **2** (6), 58-68, (2014).

Analysis of Public Service Advertisement's Creation of Fashion Brand Image

Xueying Kuang, *Shui* Jin*

Beijing Institute of fashion Technology, Beijing, China 10012

Abstract. The problem solved by this article is that the new media is increasingly showing a vigorous development, the market competition is becoming more and more fierce, media technology and media are changing rapidly, and traditional advertising can no longer meet the competitive development of some modern brands and the needs of the audience. The research process of this article is from the perspective of public service advertisements, studying the development status of domestic and foreign brand public service advertisements and the case analysis of its application in brand image shaping and dissemination. The conclusion of this article is that now that homogenization is getting higher and higher, if you want to build a brand that is recognized and loved by consumers, you must effectively distribute brand products, reflect the brand's social responsibility and love. The unique research of this article is the integration of public service advertising and fashion brands to create an innovative research model that combines practice and promotion.
Keywords: Public service advertising, Image shaping, Sense of responsibility

1 Introduction

Nowadays, new media is showing an overall trend of vigorous development, and market competition is becoming more and more fierce, the media technology and medium are changing with each passing day. Traditional advertising can no longer meet the competitive development of some modern brands and the needs of audiences. Therefore, the advent of new media era will also bring a new form of brand image shaping, Public service advertising. With the innovation and progress of science and technology, public service advertising has more development opportunities, but the increasingly complex new media environment also makes public service advertising face unprecedented difficult challenges. The corporate behavior and brand competition of fashion brands will be launched at a higher level. The scale operation and grouping trend of corporate brands are also more obvious. Brand competition has developed from resource competition, talent competition and information competition to The overall competition between enterprises-brand image competition [1]. A brand can devote itself to the interests of the society, think about it from the standpoint of the public, conscientiously shape the good public image of the brand in the society, and make it win the favor of the public and consumers in the market competition and be invincible.

* Corresponding author: jinshui@126.com

2 Public welfare and brand win-win under the media ear

2.1 The social value of public service advertisements

In the report of the 19th National Congress of the Communist Party of China, General Secretary Xi Jinping pointed out that it is necessary to cultivate and practice the core values of socialism and give play to the leading role of the core values of socialism in national education, the creation of spiritual civilization, and the creation, production and dissemination of spiritual and cultural products [2]. Especially in September 2017, the Central Committee of the Communist Party of China and the State Council issued the "Opinions on Creating a Healthy Growth Environment for Entrepreneurs and Promoting Excellent Entrepreneurship to Better Play the Role of Entrepreneurs" to further promote and promote entrepreneurship [3]. At the same time, the rapid development of society is bound to bring more support to public service advertising. The state has also given updated requirements and industry basis for public service advertising. The 2016 "Interim Measures for the Promotion and Management of Public Service Advertising" won the national business Examined and approved by the Bureau of Administration of the General Administration of Administration, issued by six ministries and commissions at the same time, which pointed out that "public welfare advertisements funded by enterprises for design, production, publication or naming can be marked with the name of the enterprise and the trademark logo". These national policies and corporate behaviors are not only related to social economic development, but the ultimate goal is to build a more harmonious society. Therefore, it can be said that there is a complementary relationship between public service advertising and social development.

2.2 Image enhancement of fashion brands

In recent years, in the face of fierce market competition, fashion brands have fought fiercely, and many brands compete for price with quality competition, resulting in a negative growth in profit income, and ultimately undoubtedly caused a huge obstacle to the development of the brand itself. At the same time, the investment in brand promotion is obviously insufficient, and it is caught in a vicious circle. The competition of modern fashion brands is no longer just a single-level competition centered on products, but has transitioned from the competition of products to the era of brand competition. In short, it is the competition of brand image.

Consumers' memory of a brand is essentially the memory of the brand's image, and consumers' choice of products is not only considering the price factor, but also depends on the consumer's inner identity. If a brand does not have a special flash point or It is a clear positioning, and it is easy to be forgotten by consumers. Brands without a personal image are not easy to exist for a long time. Under the market economy, brands are essentially products of market competition, a sign of quality, and a sign of consumer recognition. Today's market competition is actually the competition between brands [4]. Improving the brand image and sustainable competitiveness is not a one-off, one-shot thing, which is essentially a huge project that requires full consideration.

2.3 Mutual benefit and mutual benefit of public welfare and brand

With the advancement of technology, new media is showing a vigorous development trend, the competition between brands is intensifying, pure commercial advertising may bring temporary profits to corporate brands, but if you want to continue healthy development, keep the brand lasting. The vitality of commercial advertising alone will not work [5]. Traditional advertising can no longer meet the needs of the development of the modern advertising industry and the audience. It is precisely in this context that the advent of the new media era has brought a new form of advertising, Public service advertising, thus the vigorous development of the advertising industry provides a more effective communication carrier. In the invisible battlefield of fashion brand competition, its brand behavior and brand competition will be launched at a higher level, and the scale operation and grouping trend of brand companies are more obvious. Brand competition has been from resource competition, talent competition, information Competition has developed into a brand's overall competition-image competition [1]. If a brand can serve the public interest, and thus shape its own social public welfare image, gain the goodwill and recognition of the public, and resonate with the society, then it can win the good impression of public and consumers, and be invincible in the market competition.

Western countries have a long history of development in the advertising industry, and the creation system of public service advertisements is more mature and diverse. The investment and research promotion of foreign fashion brands in public service advertisements has entered a mature period. These large companies are keenly aware that although public service advertisements cannot directly promote their products, on the contrary, this mode of dissemination of weakened business information will also directly serve the interests of the public, so as to achieve better communication with consumers. Public service advertisements can greatly highlight and emphasize the brand's social responsibility, and give the company a good and noble social public image. Frequent broadcasting can also strengthen the brand impression that the brand brings to the public [5]. Indirectly brings to the brand Good reputation and many other intangible value returns. Being able to build a noble and good image of the brand in the minds of consumers is actually a way to promote the brand itself. It can be seen that the impact of public service advertisements on brands and companies seems to have no immediate improvement effect, but its long-term impact in consumers' minds and consumers' recognition of their overall brand image is any conventional model of commercial advertising could not be compared with and surpassed.

3 Public service advertisements shape the brand's own image

3.1 Embodiment of brand responsibility

In fact, the sustainable and healthy development of modern brands depends on whether they can obtain public recognition and support, and whether they have a brand image that is deeply rooted in people's hearts, and whether they can have far-reaching influence. Brands make public service advertisements to disseminate and explain various issues that arise in the development of social civilization, guide the public to pay attention to certain social issues, and change public attitudes and views on these public welfare issues [6], and urge the public to make The good things that are beneficial to society and the natural environment, supporting the advocacy of a certain social style to promote the alleviation or resolution of social problems, played a role in social education. A questionnaire survey I conducted also shows that all sectors of the community generally have a certain preference for public service advertising brands, and do not exclude public service advertising. Secondly, as the pressure of competition among brands has increased, while commercial advertising has sprung up like mushrooms, there are also some urgent problems to be solved. For example, some commercial advertisements are obviously false and deceptive, which obviously exaggerates the role and quality of their finished products. The utilitarian nature and indirect profitability of public service advertisements are in sharp contrast with commercial advertisements. Not only can they establish a good and noble social image of the brand, highlight the brand's strong sense of social responsibility and love, but also allow consumers' emotional experience and psychological satisfaction In the first place, it unconsciously strengthened the confidence in the brand, and established a good image in the public's mind in a subtle way, thus enhancing the brand's popularity and brand value.

3.2 Enhancement of brand affinity

Emotions are indispensable in our lives, but what are emotions specifically? I think it is a deep connection, the result is to make people think deeply or resonate. In today's social life, consumers' acceptance and choice of a brand is not only to satisfy the functional value of its brand, but also to be an emotional appeal [5]. As the saying goes, people who win the hearts of the people win the world, and public service advertisements can eliminate consumers' doubts in the easiest way, and then narrow the distance with consumers. Good public service advertisements can "single hit" with the inner thoughts of consumers, starting from the scenes and examples we are familiar with, letting The audience has a strong sense of resonance and achieves the purpose of long-term communication, which is also the communication effect that ordinary commercial advertisements cannot achieve.

"Advertisement is the epitome of life, showing all kinds of human beings [7]. And there is a type of advertisement that not only expresses life, but also forms a force for good, and promotes the progress of society little by little. This is public welfare. advertising". Relative to the all-star endorsement tactics frequently used by competitor Pepsi, Coca-Cola has always been more enthusiastic about digging up the most subtle warm stories to gain popularity, and chooses to communicate deeply with consumers emotionally. Coca-Cola's public service advertisements are more of an emotional and spiritual lubricant, some even focus on sensitive and heavy social issues. For example, one of their public service advertisements focusing on South Asian workers working in Dubai has successfully captured the issue of the social working class. As we know that there are a lot of workers from South Asia. In order to save every cent of their hard-earned money, they are not even willing to spend phone calls to call home, so the connection with their loved ones is brutally cut off by reality.

After the Coca-Cola company learned about this situation, they immediately invited the relevant professional team to develop a phone booth device that can use the Coca-Cola bottle cap as a call fee, so that workers in foreign countries can Listened to the voices of distant relatives at a lower telephone rate, and used this as a content to shoot public service advertisements, which won unanimous praise from the general public and inspired thousands of consumers. Curiosity and pain points have made the brand warmer and more humanize. While shaping the image of the Coca-Cola company full of social responsibility, it has also achieved good social benefits [5]. From this we can see that public service advertising is using its unique value orientation to create a more harmonious and beautiful society for us.

4 Public service advertisements are a powerful tool for brand communication

4.1 Soft ads establish closer communication

Today, with the rapid development of society, the public is becoming more and more sensitive and antipathy to hard advertisements with too much brand commercial information, and they are more and more passively accepted. In such a large environment, consumers' psychological resistance to commercial advertising is increasing day by day, and public service advertisements stand precisely from the perspective of consumers, observing the feelings and feelings of thousands of people, and appearing as a bosom friend. This approach breaks people's defense against the brand psychologically, aroused wide public resonance and goodwill, and made up for the declining trust of commercial advertising, thus winning the trust of consumers. Public service advertisements be emotional in the content and appeal of the story, and stimulate the audience's "emotional response" is undoubtedly an effective strategy. This shows that public service advertisements enjoy higher acceptance than commercial advertisements, and are easier to impress and persuade the audience.

4.2 Cumulative reputation improves audience recognition

A responsible brand will conduct behaviors that give back to the society, and brand making public service advertisements and participating in public service actions is one of the effective ways.

While consumers stop to watch the brand's high-quality public welfare advertisements seriously, the resonance that emerges from their hearts is actually a natural acceptance of the brand. The brand can use its inner emotional explosive power to win the resonance of the audience. Once the resonance is formed, the emotional effect of the advertisement will be reached, and the brand reputation will be quietly absorbed and improved.

IKEA, a well-known international home furnishing chain brand, is closely linked to the attributes of the industry and is extremely close to consumers. In their advertisements, it advocates "a healthier planet, starting at home: exploring products and solutions for a more sustainable life. " Compared with the slogan advertising of Rotten Street, "Protect the environment, everyone is responsible, protect the environment, and start with me", this kind of communication that closely follows the value of the brand, the former pays more attention to humanized communication with consumers and is more humane, creating a warm, harmonious and comfortable atmosphere and imagination space. This advertisement also properly spreads the business philosophy of the IKEA brand, and the appeal of the advertisement and the connotation of the brand have been cleverly integrated. This act of protecting the environment from its own product actions has shaped the image of an IKEA company full of social responsibility, while also accumulating public reputation for the brand. And the way of emotional persuasion makes it easier for consumers to have a good impression on the brand, so that consumers can readily accept the choice of furniture decoration at home can also contribute to the protection of the global environment, thereby improving the audience's sense of identity and brand loyalty, and achieving good Social benefits.

After the brand gains the trust of consumers, consumers will have a tendency to "affirm analogies." In the future consumption, the brand's products or services will be implicitly considered as the first consideration, thereby increasing brand loyalty. The rapid development of the economy has also promoted the diversification of commodities, and there are countless homogenous commodities, and in the face of this situation, consumers will naturally choose those familiar and well-publicized brand commodities.

4.3 Promote public welfare as a role model

Fashion brands are not only be the fighters in the fierce market competition, but also the promoters of society towards harmony and beauty, assuming their due social responsibilities. The mistakes and deviations in the direction of social ethos and values may affect the moral system of the entire society, and will also hinder the harmonious and stable development of society [8].

Public service advertising is a mirror that reflects society. It is characterized by a wide range of audiences, strong penetration and strong themes. It is a reliable communication carrier for the social masses to spread mainstream social values, and to promote and advocate to the public through the form of advertising These positive and positive messages. On the one hand, brands can integrate real-time social hot issues or ideas that the brand wants to promote into the public service advertisement through a unique performance method, which can resonate with the public and achieve publicity and advocacy.

On the other hand, in this large environment, everyone's own behavioral norms will be subtly influenced and constrained by this positive and positive energy environment, no matter in which class or environment, Such positive ideas will be vigorously spread through public service advertisements, [8] when the masses will change in a positive direction, which can enable everyone in the society to form a good moral concept system, and then regulate themselves. The process of audience watching public service ads is actually the process of accepting their content ideas and value orientation. [8] The ancients said that "the person who is close to Zhu is red, and the person who is close to Mo is black". That means even an uneducated personality can be influenced and educated by a large number of outstanding public service advertisements, so that everyone can continue Study and progress in order to improve and improve one's own moral behavior and jointly promote social development and progress.

For example, the National Bank of Dubai ENBD (one of the largest banks in the UAE) launched a heart-warming public service advertisement on November 13 to commemorate World Good Day [9]. This public service advertisement tells the story of the indifferent selfish male character who is finally "infected" by kindness. In the advertisement, the male master actually insinuates everyone in real life who refuses to help. But in fact, as long as you are willing to try to help others once, you will feel the joy of giving. Humans are social animals that are easily affected by infection. A smile, an action, and a help are likely to embark on a new life track. On the day of International Friendship Day, ENBD used this public service advertisement to depict the power of help. Once this public service advertisement was broadcast, it caused widespread concern and deep reflection from the public. It is telling us that "you are probably the one who helped him take off his indifferent coat". While advocating mutual help, "coldness" should not be synonymous with people in contemporary society[10].

5 Conclusion

5.1 Integration of brand and public welfare becomes possible

With the continuous development and maturity of science and technology, new media and new technologies are rapidly entering all walks of life, and the advent of the Internet era and the emergence of various new media have provided various effective ways for the dissemination of public service advertisements. Public service advertisements require innovative forms of communication, augmented reality interactive technology, and the use of innovations in communication forms and the enhancement of interactive participation is the key to the effect of public advertisements in stimulating widespread communication. [11] One of the biggest features of the new media era is mass media-covering the vast majority of online media users, especially with the popularization of mobile phone short video software. Brands making public service advertisements should be more compatible with the characteristics of mobile phones, fit the communication methods of mass media, and make use of the interactivity and immediacy of mobile networks to spread widely.

Secondly, with the help of new media and new technology forms, improving the interactive participation of public service advertisements can make up for the shortcomings of the traditional one-way communication of public service advertisements. [12] For example, the "VR + Public Service Advertisement" model can effectively stimulate the curiosity and participation enthusiasm of the audience, provide the audience with immersive feelings, enhance interactivity and satisfy the emotional desire of the audience, so as to achieve the effect of shocking the soul and effectively resonate with the audience , Which enhances the interest, interactivity, communication power and influence of public service advertisements, making the effect of advertising delivery more significant.

The more important thing is that the shooting of public service advertisements is indispensable for the image building of fashion brands, and it is even more effective than traditional propaganda. [13] Public service advertisements have unique advantages such as wide audience groups, resonating with the public, diverse communication carriers, high arrival rate, high transmission rate, and wide communication channels. They can stimulate the inner resonance of the audience to spontaneously spread and form a good social effect. At the same time, it creates a good positive brand image in a wider range and at a deeper level, so that the brand can continue to develop healthily.

Acknowledgments

This work was supported by the Special Fund for the Construction of High-level Teacher's Group of Beijing Institute of Fashion Technology, Project No.biftxz201905.

References

1. G Chao. The development status and countermeasures of public service advertising in my country. News enthusiasts (the second half of the month), (2010).
2. C Qi. An Analysis of the Path of Socialist Core Values Leading the Construction of Campus Culture in Universities——Taking Hengyang Teachers College as an Example. Knowledge Economy, **23**, 44+46, (2018).
3. J Yin. Promoting the spirit of excellent entrepreneurs to better play the role of entrepreneurs. Economic Management: Digest Edition.
4. Q Li. Research on the Construction of Regional Tourism Brands—Taking Dahuangshan Tourism District as an Example, (2009).
5. H Qin. On public service advertising and corporate brand image building. News Window, **04**, 55-56, (2014).
6. D Xu, C Wang. An Analysis of the Cultural Connotation of the Chinese Characters in the Water Department of "Commonly Used Character Tables in Modern Chinese". Journal of Liaoning Normal University (Social Science Edition), **6,** (2016)
7. Chinese advertising magazine. Santa also has no place to give gifts.
8. http://www.sohu.com/a/278720379_657211,2018-11-30
9. S Lin. Analysis of the characteristics of public service advertisements and their ideological and political education functions. Xuelun, **12,** 178-179+186, (2018)
10. Y Wang. Analysis of the style of campus public welfare advertisements. Young writers, **21,** 95-95, (2010)
11. Advertisers_discard.2018 Top Ten Public Service Advertisements.
12. http://www.sohu.com/a/288289192_117194, (2019)
13. F Qin, Z Gu. Pathfinder's brand communicateon research, China Sports Industry and Sports Goods Industry Development Forum, **0,** (2018)

14. J Zhang, M Hu. Research on the development trend of the integration of VR technology and public welfare communication [J]. Sound Screen World· Advertiser, **04**:125-128, (2019)
15. O'Barr W M. Public Service Advertising. Advertising & Society Review, **7**(2):1597-1597, (2006)

Design of Rehabilitation Training System for Pelvic Floor Dysfunction

Yongfeng Liu[1], *Chunna* Li[2], *Jun* Zhong[1*], *Liming* Cai1 and *Kai* Guo

[1] Suzhou Institute of Biomedical Engineering and Technology, Chinese Academy of Science, Suzhou, Jiangsu, 215163, China.
[2] 964 Hospital of the Joint Service Support Force of the Chinese People's Liberation Army, Changchun 130062, China.

Abstract. Pelvic floor dysfunction has caused pain to the lives of the majority of female patients, and repeated visits to the hospital for treatment have also caused inconvenience to patients. Therefore, a portable pelvic floor dysfunction rehabilitation training system is designed to help female patients with personalized and private treatment. The pelvic floor dysfunction rehabilitation system realizes the independent training mode selection through the APP client, assists the pelvic floor muscle rehabilitation training through the electrical stimulation module, and can realize the recovery situation of the pelvic floor muscle strength in real time through the AD acquisition module. The experimental results show that the pelvic floor dysfunction rehabilitation training system implements bipolar electrical stimulation pulses, and the EMG signals collected at the same time clearly show the EMG signal strength and strength maintenance time.

1 Introduction

Childbirth and pregnancy activities, abdominal surgery, etc [1-3], will cause varying degrees of pelvic floor function reduction or obstacles, which may be accompanied by lower abdominal discomfort, urinary incontinence, abnormal urination, abnormal bowel movements, abnormal vaginal secretions, etc., which will bring female patients later life Inconvenience and pain, taking pelvic floor muscle rehabilitation training in time for childbirth or after surgery is of great benefit in preventing postpartum complications and restoring pelvic floor muscle function [4, 5].

At present, many studies have shown that pelvic floor electrical stimulation has obvious effects on pelvic floor muscle rehabilitation [6-8], but due to the large hospital treatment equipment and long treatment cycle, it is not convenient for female patients to complete the treatment course better. In order to enable female patients to carry out early and timely intervention on pelvic floor dysfunction at home, it is designed a rehabilitation training system for pelvic floor dysfunction.

The pelvic floor dysfunction rehabilitation training system designed in this paper uses the constant current electrical stimulation module of the pelvic floor dysfunction rehabilitation

* Corresponding author: zhongj@sibet.ac.cn

training device to electrically stimulate the pelvic floor muscles, and uses the AD module to collect pelvic floor muscle strength signals and send the EMG signals through Bluetooth Go to the mobile APP for display and management. The patient uses the APP to view the current pelvic floor muscle signal strength and pelvic floor muscle recovery. This pelvic floor dysfunction rehabilitation training system has the advantages of high integration, low power consumption, high measurement accuracy, etc., which is convenient for female patients to use and relieve the pain caused by pelvic floor dysfunction.

2 Overall System Design

The pelvic floor dysfunction rehabilitation training system includes pelvic floor dysfunction rehabilitation training apparatus and APP. The pelvic floor dysfunction rehabilitation training device includes a main control MCU module, a constant current electrical stimulation module, an AD acquisition module, a Bluetooth module and a power management module, and the APP part displays and manages training data. The functional block diagram of this system is shown in figure 1.

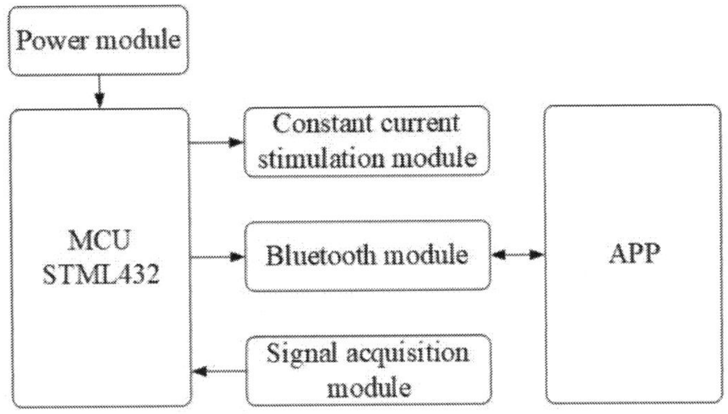

Fig. 1. System block diagram.

3 Selection of Main Components

3.1 Main Control Chip STM32L432

STM32L432 is an ultra-low power microcontroller based on 32-bit RISC core, and its operating frequency can reach 80MHz. The Cortex-M4 core has a floating-point arithmetic unit that supports all Arm® single-precision data processing instructions and data types. This chip provides a 12-bit ADC, the speed can reach 5Msps, while also providing low-power RTC and timer to reduce power consumption.

3.2 Bluetooth Chip NRF51822

nRF51822 is a powerful and highly flexible multi-protocol integrated circuit chip, commonly used in 2.4 GHz Bluetooth low energy applications. nRF51822 has 256kBFlash and 16kBRAM, and also has a wealth of analog and digital peripheral products. It can interact through the programmable peripheral product interconnection (PPI) system without CPU involvement, and can flexibly design the PCB through the pin mapping scheme. It is now used in medical, fitness, consumer electronics and many other fields.

3.3 AD Acquisition Chip ADS1292

ADS1292 is a multi-channel 24-bit delta-sigma analog-to-digital converter with built-in programmable gain amplifier, internal reference and an on-board oscillator. Its operating data rate is up to 8KSPS, and 335uW per channel in low power mode.

4 System Hardware Design

4.1 Constant Current Electrical Stimulation Module

The constant current electrical stimulation module consists of a booster circuit and an H-bridge circuit, as shown in figure 2. According to the set stimulation current value, STIM_PWM1 transmits a 30 KHz frequency PWM wave, and the duty ratio corresponds to the set current value. Through the charge pump circuit composed of L3, D2 and Q5, the input 3.3V is pumped to a high voltage (10-55V) and rectified by C28. Q6 is designed as a protection circuit. When the detection point STIM_V2 detects that the pump voltage exceeds the target voltage, the circuit is discharged through Q6. At this point, the circuit completes the boosting process and boosts the 3.3 V voltage to a constant voltage STIM_VMAX output.

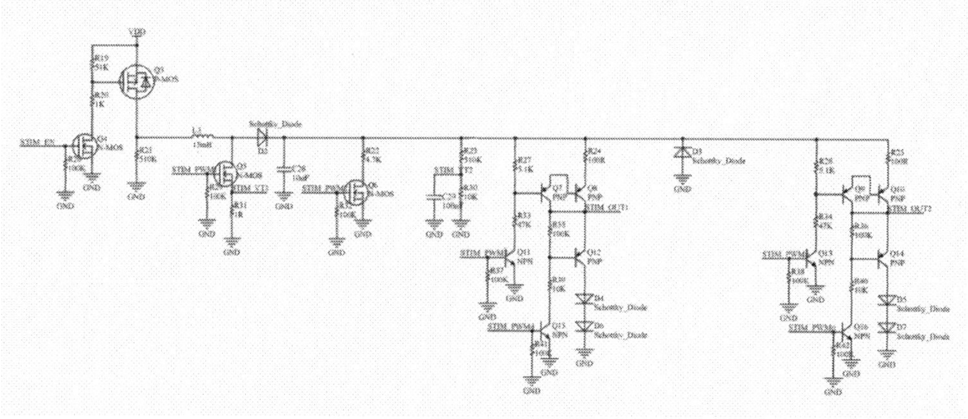

Fig. 2. Circuit diagram of constant current electrical stimulation.

The stimulation pulse waveform generated by the H-bridge circuit in the system is a bipolar pulse. The positive and negative pulse charges are equal. The positive and negative charges neutralize each other in a cycle, which can reduce the accumulation of charges in the skin tissue, which is not easy to cause muscle fatigue and can be avoided by Unipolar pulse stimulation causes problems such as a tingling sensation due to charge accumulation. When the system is working, a constant current circuit is formed by Q7, Q8, Q11, Q15, R27, R33, R35, R39, and the second half is composed of another constant current circuit, and positive and negative stimulation is performed through the form of H bridge: STIM_PWM3 Flat, STIM_PWM4 low level, STIM_PWM5 low level, STIM_PWM6 high level, at this time STIM_OUT1 level is higher than STIM_OUT2, positive input; when the control level is all reverse, the corresponding input is negative.

4.2 AD Acquisition Module

ADS1292 selects 1-channel differential signal input, and after 24-bit analog-to-digital conversion, data transmission is performed through the SPI interface of the MCU.

4.3 Bluetooth Data Transmission Module

The serial port of NRF51822 is connected to the serial port of MCU. The APP on the mobile phone sends the training mode command via Bluetooth. The data processed by the MCU is sent to the Bluetooth module through the serial port, and then sent to the APP on the mobile phone via Bluetooth for display and management.

5 System Software Design

5.1 Control Software Design of Rehabilitation Training Apparatus for Pelvic Floor Dysfunction

The control program of the pelvic floor dysfunction rehabilitation training instrument includes two parts: the main control program based on STM32L432 and the Bluetooth data upload program based on NRF51822. Keil uVision5 is used to write STM32 embedded program and Bluetooth data upload program. The entire drive control flow of STM32 is shown in figure 3. After the system is turned on, initialize the timer, SPI, PWM, serial port, etc., wait for the mobile phone APP to send training instructions via Bluetooth, and after receiving the instructions, set the frequency, pulse width and other parameters of the constant current electrical stimulation module according to the mode selected by the instruction, and Start the electrical stimulation training, and at the same time use the AD acquisition module to collect pelvic floor muscle strength signals. STM32 sends the EMG signal data to the Bluetooth module NRF51822 through the serial port. After receiving the data, the Bluetooth module NRF51822 sends it to the mobile phone APP via Bluetooth for display and management.

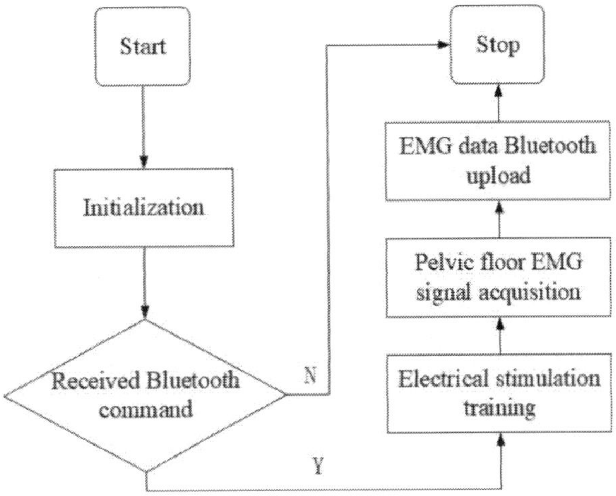

Fig. 3. Embedded drive control flow chart.

5.2 APP Design of Pelvic Floor Dysfunction Rehabilitation Training System

The pelvic floor dysfunction rehabilitation training system APP provides patients with the choice of treatment mode. At the same time, the pelvic floor dysfunction rehabilitation training device uploads the pelvic floor myoelectric signal to the mobile phone APP for display. The patient can view the strength of the pelvic floor muscle strength signal through the APP By comparing with historical data, you can view the recovery trend, so that patients have a clearer understanding of their pelvic floor function recovery.

6 Experimental Test

6.1 Simulated Electrical Stimulation Experiment

In the experiment, a $1K\Omega$ resistor was used to simulate the human body impedance. The output of the electrical stimulation module was connected to both ends of the resistor through the electrode pads. The electrical stimulation parameters were set for electrical stimulation output. The actual waveform was observed through an oscilloscope. The experimental results are shown in figure 4. From the experimental results, it can be seen that the output waveform amplitude is 16V, the stimulation frequency is 35Hz, and the pulse width is 125uS, which meets the design requirements of electrical stimulation therapy.

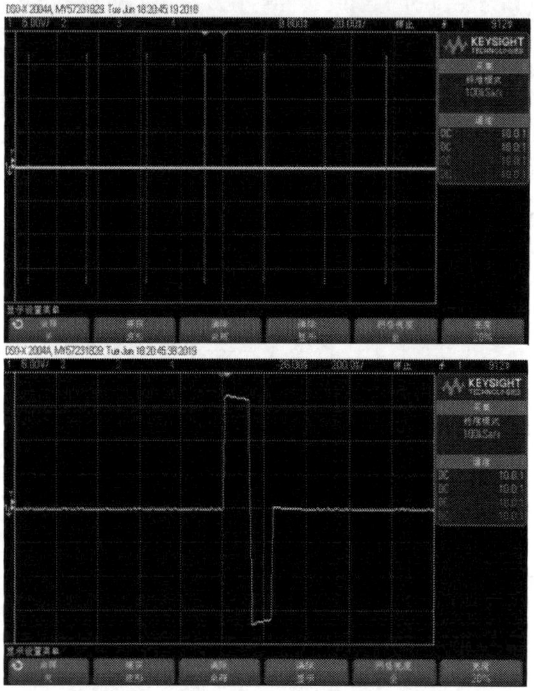

Fig. 4. Electrical stimulation signal diagram.

6.2 EMG Signal Acquisition Experiment

In the experiment, the electromyographic signal of the human biceps was connected to the input terminal of the AD acquisition module through the electrode sheet, and the 5 kg weight was lifted by the arm every 5 seconds to generate the electromyographic signal of the biceps, and then the weight was dropped for 5 seconds Repeat the test for 1 minute, the test results are shown in figure 5. It can be clearly seen from the figure that the muscle EMG signal becomes stronger when the muscle is exerted, and during the time period of maintaining the strength, the EMG signal is basically maintained at about 100 μ V-150 μ V. The strength and duration of EMG signals are both important indicators for evaluating the recovery of pelvic floor muscle strength.

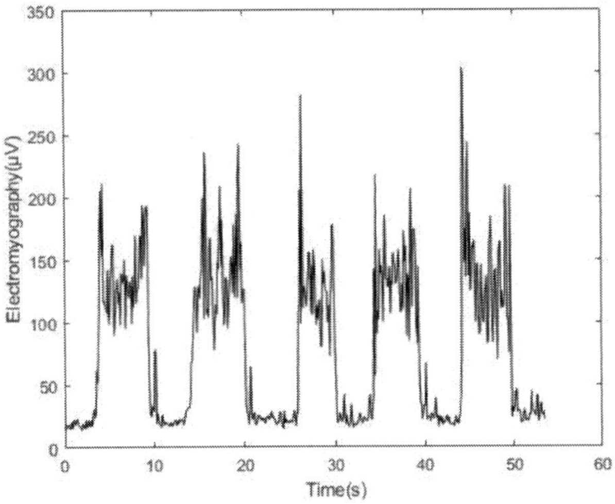

Fig. 5. EMG signal acquisition diagram.

7 Conclusion

The pelvic floor dysfunction rehabilitation training system includes pelvic floor dysfunction rehabilitation training instrument and APP client. It has the functions of large-scale hospital equipment and is also miniaturized and portable, not limited by time and space. It is used in various occasions and has three major functions: electrical stimulation, muscle strength test and biofeedback autonomous training. Induce the electrical activity of the pelvic floor muscles of the patient through the vaginal probe to evaluate the strength of the pelvic floor muscles, and then start the electrical stimulation program to stimulate the pelvic floor muscles and their inner nerves with electric current to directly induce a therapeutic response, which can be a good solution for women Patients with pelvic floor dysfunction [9, 10].

Acknowledgements

This work was supported by "Jiangsu Science and Technology Plan: Construction and Research of Rehabilitation Treatment and Information Management Platform for Pelvic Floor Dysfunction" (BE2017672).

References

1. W Y Du, Y F Zhou, Y T Yuan, etc. *The effect of Kegel exercise combined with pelvic floor rehabilitation on pelvic floor function in patients undergoing laparoscopic extrafascial hysterectomy*, Journal of Navy Medicine, **42** (2) 203-205(2020)
2. F F Jia, *Analysis of the effect of postpartum pelvic floor rehabilitation exercise combined with bioelectric stimulation on the postpartum pelvic floor function rehabilitation effect and prognosis*, Modern Chinese Medicine Application, **14** (7) 226-227 (2020)

3. LM Yuan, L Li, D X Liu, etc. *Rehabilitation effects of pelvic floor muscle training on female pelvic floor dysfunction diseases*, Modern Medicine and Health, 36, 6: 918-919 (2020)
4. I MILSOM, *Can we predict and prevent pelvic floor dys-function?* Int Urogynecol J, **26**(12) 1719-1723 (2015)
5. K JUNDT, U PESCHERS, H KENTENICH, *The investigation and treatment of female pelvic floor dysfunction*, Dtsch Arztebl Int,**112**(33/34) 564-574 (2015)
6. L S Liu, *Application of pelvic floor muscle training and electrical stimulation in postpartum pelvic floor muscle rehabilitation*, Journal of Mathematical Medicine, **33**(4) 606-607 (2020)
7. C H Wang, *Pelvic floor rehabilitation training physiotherapy instrument in patients with pelvic floor organ prolapse application effect*. Medical Equipment, **33**(6) 153-155 (2020)
8. T R Wang, X Zhou, L J Jiang, *The effect of comprehensive treatment of pelvic floor and its influence on quality of life of urinary incontinence in the elderly women*. Chin J Clin Healthc, **23**(2) 252-255 (2020)
9. S J MADILL, S P DRILET S, A TANG A, etc. *Changes in urethral sphincter size following rehabilitation in older women with stress urinary incontinence*, Int Urogynecol J, **26**(2) 277-283 (2015)
10. A RADZIMINSKA, A STRACZYNSKA, M W RAJEK, etc, The impact of pelvic floor muscle training on the quality of life of women with urinary incontinence: a systematic literature review. ClinInterv Aging, **13**(5) 957-965 (2018)

Research on the Efficiency and Potential of China's Trade with South Asian Countries-Based on the Stochastic Frontier Gravity Model

Yifan Lei[1*], *Cuiping* Li[2]

[1]Joint training of Xinjiang University of Finance and Economics and Central University of Finance and Economics， Institute of Finance, China, e-mail: 1615877288@qq.com
[2]Ptgraduate tutor, School of Xinjiang University of Finance and Economic, China

Abstract. In the context of COVID-19, international trade has been strongly impacted, and it is of great significance to accelerate trade exchanges with countries along the "Belt and Road" to achieve trade transformation. This paper uses the time-varying stochastic frontier gravity model to measure trade efficiency based on the data of China and South Asia from 2000 to 2019. The empirical results show that economic scale and population have a positive impact on trade. China's GDP growth is not conducive to the growth of bilateral trade volume, and distance is no longer a factor hindering trade. Trade inefficiency has a great impact on trade. Among them, whether to sign a free trade agreement, the quality of port infrastructure, the simple average tax rate of all products, and political risks have a more significant impact on the trade coefficient. The trade potential between China and South Asian countries is very large and different countries are uneven. China has high trade efficiency with Sri Lanka, Pakistan, India, and Bangladesh, while trade with Bhutan has extremely low efficiency, and trade with Maldives and Nepal is average. Based on the above empirical results and combined with the current epidemic situation, the corresponding conclusions and countermeasures are put forward. For example, exporters should continue to innovate to promote trade transformation and diversified development; combine the western development strategy to promote trade and South Asian exchanges; reduce trade tariff barriers between the two countries; Jointly maintain the security and political situation of the two countries.
Keyword: South Asia, Trade Efficiency, Potential, Stochastic Frontier Gravity Model

* Corresponding author: cdjrlcp@126.com

1 Introduction and literature review

1.1 Introduction

In 2020, the new epidemic will sweep the world, and international political relations will undergo subtle changes. South Asia is an important area of the "Belt and Road" and "Maritime Silk Road", and has also established the China-Pakistan Economic Corridor and the Bangladesh-China-India Economic Corridor with China to promote the development of the two countries. During the epidemic, trade exchanges between China and South Asian countries were restricted. On November 11, 2020, in order to speed up the resumption of work and production, China and South Asia held an international trade digital exhibition. The common interests of China and South Asian countries are accelerating and the potential for economic and trade exchanges continues is deepening, enterprises are increasingly eager to strengthen mutual benefit and mutual assistance. The China Business Council will continue to work hand in hand with other countries to create a better platform for China and South Asian countries to carry out trade activities, provide quality services, and create convenience. It shows that the two sides have strong hopes to jointly strengthen the trade between the two countries, and it is of great significance to continue to deepen the friendship and trade exchanges between the two parties in the context of major emergencies.
From 2005 to 2018, China's exports to South Asia increased by about 4.5 times. The trade exchanges between the two sides have become increasingly close, but the growth rate is still very low, and the two sides have great potential for development. The "India and South Asia Limited Trade Links" report confirms that South Asia accounts for 5% of the national trade, and the level of integration of South Asian countries is significantly lower than the level of global economic integration. The reasons include trade protectionism, high logistics and transportation costs, lack of political will, and widespread trust deficits. In order to enhance the trade potential of bilateral countries, this article conducts empirical analysis and exploration and gives corresponding opinions and policies.

1.2 Literature review

Regarding the study of trade potential, most scholars use SFGM to measure trade efficiency and then analyze trade potential. Research on foreign scholars: Amirul Islam, Ruhul Salim and Harry Bloch (2016) [1] studied the impact of free trade agreement signing on South Asia's efficiency and productivity growth. The results show that South Asia still has the problem of declining production efficiency and pointed out that backward countries may increase productivity by moving resources across borders. Rakesh Kumar (2018) [2] estimated the current situation of South Asian trade integration under the self-returning multilateral framework, and the results showed that Pakistan and Sri Lanka have long-term trade complementarity, and India and Bangladesh, India and Sri Lanka have short-term trade complementarity. Nazia Gul and Hafiz M. Yasin (2011) [3] used a trade gravity model to estimate Pakistan's trade potential. The results showed that Pakistan has the largest trade potential with countries in the Asia-Pacific region, the Middle East, Latin America and North America, and its trade volume with South Asia is very low. The main reason is political tensions in neighboring countries. Naseem Akhter and Ejaz Ghani (2010) [4] used the gravity model to assess the trade potential between the member states and non-member states of the South Asia Free Trade Agreement. The results showed that the trade volume will increase only when the major partners sign trade agreements. Research by domestic scholars: Yingying Zong (2018) [5] studied the complementarity of agricultural product trade and export similarity between China and South Asian countries and proposed corresponding countermeasures. Zhixiong Liu and Honghui Chen (2020) [6] found from static and dynamic empirical research that China's trade with South Asia is mainly inter-industry trade and there are no large country differences. Yunjie Cheng and Chenghui Dong (2020) [7] used a heterogeneous stochastic frontier gravity model to measure the efficiency of bilateral trade. Jin Liu (2020) [8] explored the scale and structure of trade between China and South Asia. Ling Wang and Shan Chen (2019) [9] studied the trade efficiency of the Bangladesh-China-India Economic Corridor and found that trade efficiency has been declining over time. They proposed to speed up free trade zone negotiations and reduce trade barriers. Zhigang Gao and Feng Tian (2019) [10] Study the trade efficiency of the China-Pakistan Economic Corridor. Liang Li and Shucai Ma (2020) [11] research on trade competition and complementarity between China and South Asia shows that their country has advantages in industrial products and South Asia has advantages in low-end products. The two countries have a strong complementary relationship.

2 Model settings, variable descriptions and data sources

2.1 Theoretical model

2.1.1Trade gravity model (GM)

The gravity model originated from Newton's universal gravitation. Tinbergen (1962) and Poyhonen (1963) first used GM to measure trade efficiency, and its basic form is as follows:

$$T_{ij}=\frac{A(GDP_i GDP_j)}{D_{ij}} \tag{1}$$

T_{ij} represents the volume of trade between the two countries, A is a constant, GDP_i and GDP_j represent the gross domestic product of the traded countries respectively, and the volume of trade is positively correlated with GDP, D_{ij} represents the distance between the two countries, and the volume of trade has a negative correlation with the distance .

In the empirical analysis of the model, in order to explain the meaning of the coefficients, the logarithm of the data is usually taken. The basic form of the model is as follows:

$$LNT_{ij}=\alpha_0+\alpha_1 LNGDP_i+\alpha_2 LNGDP_j+\alpha_3 LND_{ij}+\alpha_4\delta_{ij}+\varepsilon_{ij} \tag{2}$$

δ is the other influencing variables, ε is the iid random disturbance term, α_0 is the constant term, and α_1, α_2, α_3, and α_4 are the corresponding variable coefficients.

Lineman (1966) added demographic variables to the trade gravity model. In subsequent research, each scholar introduced the required variables according to his own research direction to analyze the impact and measure trade efficiency.

2.1.2 Stochastic frontier gravity model (SFGM)

SFGM is a method of introducing stochastic frontiers in trade GM, which was first proposed by Meeusen and Broeck (1977). SFGM splits the random disturbance term (ε) into a random error term (v_{ij}) and a human-factor inefficiency term (u_{ij}). The trade potential obtained by the SFGM is more in line with the actual situation and the result is more effective. It is expressed as follows:

$$\begin{aligned} T_{ijt} &= f(x_{ijt},\beta)\exp(\varepsilon_{ijt}) \\ &= f(x_{ijt},\beta)\exp(v_{ijt}-u_{ijt}),\ u_{ijt}\ge 0 \end{aligned} \tag{3}$$

Take the logarithm of both sides together to get the following form:

$$LNT_{ijt} = LNf(x_{ijt},\beta) + v_{ijt} - u_{ijt}, u_{ijt} \geq 0 \tag{4}$$

T_{ijt} represents the trade volume of country i to country j in period t, x_{ijt} is a relevant variable affecting trade, β is a parameter to be estimated, v_{ijt} is a random error term, and u_{ijt} is an inefficiency term.

Calculate the trade potential, that is, when there is no human factor, the formula is as follows:

$$T_{ijt}^{*} = f(x_{ijt},\beta)\exp(v_{ijt}) \tag{5}$$

Trade efficiency is the ratio of real trade volume to trade potential. The formula is:

$$\mathrm{TE}_{ijt} = \mathrm{T}_{ijt} / T_{ijt}^{*} = \exp(-u_{ijt}) \quad u_{ijt} \geq 0, \quad \mathrm{TE}_{ijt} \in (0,1) \tag{6}$$

With the growth of T_{ijt}, the efficiency of trade also increases.

2.1.3 Trade inefficiency model (BC95)

For a long time, the estimation of the stochastic frontier is divided into two processes, that is, first use SFGM to obtain the inefficiency value, and use the inefficiency as the dependent variable for regression. However, the two stages have inconsistent assumptions about inefficiency, so it is impossible to obtain the same effective results as the first stage. In the first stage, it is estimated that the SFGM will return to trade inefficiency at the same time, adding the analysis of allocation efficiency. Kumbhakar, Ghosh and McGukin (1991) believe that the inefficiency term can represent a deterministic function of specific variables and random errors. Battese and Coelli (1995) proposed SFGM (BC95) for solving technical efficiency. The basic form is as follows:

$$u_{ijt} = \{\exp[-\eta(t-T)]\} u_{ij} \tag{7}$$

η is the parameter to be estimated considering the time-varying nature, and u_{ijt} obeys the truncated normal distribution. When $\eta>0$, $\eta<0$, $\eta=0$, the inefficiency terms decrease, increase, and remain the same with time.

$$u_{ijt} = \alpha z_{ijt} + \varepsilon_{ijt} \tag{8}$$

α is a parameter to be estimated, z_{ijt} represents an exogenous variable that trade inefficiency in period t, and ε_{ijt} is a random disturbance term.

Substituting into equation (4), the time-varying SFGM for measuring technical efficiency can be obtained:

$$LNf(x_{ijt},\beta) + v_{ijt} - (\alpha z_{ijt} + \varepsilon_{ijt}) \tag{9}$$

2.2 Model setting and variable description

Armstrong (2007) proposed that SFGM only includes core explanatory variables, such as economic scale, distance between the two countries, borders, language and other natural factors that are difficult to change in the near term; human factors such as FTA, tax rates, and freedom indicators are included in the model's trade inefficiency items. At the same time, considering that the current political situation is complex and changeable, this article adds political risk indicators as the factors affecting trade inefficiency.

2.2.1 Time-varying stochastic frontier gravity model setting

According to the above analysis, the random frontier gravity model set in this article is introduced:

$$\ln E_{ijt} = \beta_0 + \beta_1 LnPGDP_{it} + \beta_2 LnPGDP_{jt} + \beta_3 LnPOP_{it} + \beta_4 LnPOP_{jt} + \beta_5 LnDIS_{ij} + \beta_6 BOR_{ij} + v_{ijt} - u_{ijt} \quad (10)$$

β_0 is a constant term, β_1, β_2, β_3, β_4, β_5, and β_6 are parameters to be estimated, and the remaining variables are as follows:

E_{ijt} represents China's trade exports to South Asian countries in period t, and the data comes from UN Comtrade.

$PGDP_{it}$ and $PGDP_{jt}$ respectively represent China's per capita gdp in period t and South Asian countries' per capita gdp in period t,The higher the economic scale, the greater the flow between trade. The data comes from the World Bank.

POP_{it} and POP_{jt} respectively represent the total population of China in period t and the total population of South Asian countries in period t. The larger the population of China, the higher the market consumer demand and the greater the export supply. The greater the population of South Asian countries, the higher the poverty level, and the reduction in trade import demand. The data comes from the World Bank.

DIS_{it} indicates the distance between China and the capitals of South Asian countries. The farther the distance, the higher the transportation cost required, which is negatively correlated with the export trade volume. The data comes from CEPII.

BOR_{ij} is a dummy variable, which represents whether China and South Asian countries have the same borders. One is taken as 1 and no one is taken as 0. The neighboring two countries are positively related to export trade due to factors such as culture and distance. The data comes from CEPII.

2.2.2 Trade inefficiency model setting

This paper selects inefficiency variables based on the research of different scholars on trade inefficiency and the reality of China and South Asian countries. The model is expressed as follows:

$$u_{ijt} = \alpha_0 + \alpha_1 FTA_{ijt} + \alpha_2 AHS_{jt} + \alpha_3 GI_{jt} + \alpha_4 TF_{jt} + \alpha_5 FF_{jt} + \alpha_6 WEF_{jt} + \alpha_7 PR_{jt} + \varepsilon_{ijt} \quad (11)$$

α_0 is a constant term, α_1, α_2, α_3, α_4, α_5, and α_6 are parameters to be estimated, and the remaining variables are as follows:

FTA_{ijt} is a dummy variable, which represents whether China and South Asian countries have signed a free trade agreement in year t. It is 1 and 0 is not. The signing of a free trade agreement will increase the trade between the two countries and therefore has a negative effect on inefficiency. The data comes from China FTA Network.

AHS_{jt} is the simple average tax rate of all products in South Asia in period t. The higher the tax rate, the less trade between the two countries, which is positively correlated with inefficiency. The data comes from the World Bank.

GI_{jt} is the government credibility of a South Asian country in period t. The higher the credibility, the better the credibility of the country and the increase in trade exchanges between countries, which is negatively related to inefficiency. Data from the Heritage Foundation.

TF_{jt} represents the degree of trade freedom of South Asian countries in the t period. The higher the indicator, the less the South Asian countries are bound by tariffs and non-tariff barriers, which is negatively related to trade inefficiency. Data comes from Heritage Foundation.

FF_{jt} represents the financial freedom of South Asian countries in period t. The higher the index, the larger the scale of the country's financial market opening, the low degree of government control over financial services, and the improvement of financial service companies' operating capabilities, which are negatively related to trade inefficiencies. Data source Heritage Foundation

WEF_{jt} represents the quality of port infrastructure in South Asian countries during the t period (1=very underdeveloped, 7=highly developed and efficient in accordance with international standards). The higher the quality of port infrastructure, the higher the efficiency of trade. It is negatively correlated with non-efficiency items. Data source World Bank.

PR_{jt} represents the political risk index of South Asian ICRG countries in period t. The higher the risk index of a country, the lower the degree of political risk in the country, which is positively correlated with trade inefficiency. Data source PRS.

3 Empirical analysis

This article uses data from China and 7 South Asian countries (India, Pakistan, Sri Lanka, Nepal, Bhutan, Bangladesh, and Maldives) from 2000 to 2019 as a sample to study the potential of bilateral trade. The software used in this article is Frontier4.1, which is passed The LR test separately examines the rationality and applicability of the time-varying SFGM and trade inefficiency models, and makes corresponding trade potential analysis.

3.1 Model verification

To test the applicability of the model, run the LR value based on constrained and unconstrained conditions, and compare with the mixed chi-square critical value to check whether there is a trade inefficiency term and whether it is a time-varying model. The results are as shown in table 1.

Table 1. Applicability test results of the random frontier gravity model.

Null Hypothesis	Constraint Model	Unconstrained Model	LR	Degree of Freedom	10% Threshold	Test Result
There is no inefficiency	-114.70	-111.80	5.79	3	5.528	Refuse
Inefficiency does not change with time	-114.70	-112.17	5.05	2	3.808	Refuse

It can be seen from the table that the two hypotheses LR are both greater than the 10% critical value. The null hypothesis is rejected, indicating that there is a trade inefficiency term and it is a time-varying SFGM.

3.2 Analysis of SFGM regression results

This article directly makes a regression analysis on the time-varying SFGM. According to the results of Frontier regression analysis, it is shown in Table 2

Table 2. Measurement results of Stochastic Frontier Gravity Model.

	Time-varying model		
Variable	Coefficient	Standard-Error	T-ratio
Constant	-346.7466***	1.6326	-213.0069
$LNGDP_{it}$	0.7972***	0.0765	10.4214
$LNGDP_{jt}$	-0.4390***	0.1178	-3.7286
$LNPOP_{it}$	15.4658***	0.1070	114.4913
$LNPOP_{jt}$	0.9567***	0.0226	42.2419
$LNDIS_{ij}$	2.9612***	0.2595	11.4123
BOR_{ij}	-0.7218***	0.08486	-8.5052
σ^2	0.1551	13.2631	1.1694
γ	0.9953***	0.0049	204.9877
Log likelihood function	-99.0975		
LR Test	31.1965		

Note: ***, **, * represent significant at the level of 1%, 5%, and 10% respectively.

From the regression results, the SFGM coefficients all pass the 1% significance level. γ represents the proportion of trade inefficiency in the random disturbance item. γ is significant at the 1% level. The value of γ is 0.9953, which represents 99.53% of the trade resistance caused by trade inefficiency. Very close to 1 shows that human factors have an important influence on trade, and also shows the applicability of the time-varying SFGM, The following will compare the regression results with expectations:

(1) Economic scale. From the results, the GDP of both countries passed the 1% significance test. The GDP growth of South Asian countries has increased China's export trade to South Asian countries and promoted the development of bilateral trade. The correlation coefficient is 0.7972, indicating that the GDP growth of South Asian countries has a significant impact on the trade between the two countries, which is consistent with the expected result, while China's GDP growth actually reduces the trade between the two countries, inconsistent with the expected results. The reason may be that with the adjustment of the industrial structure in recent years, China's exports have shifted from agricultural products, chemical products, and textiles to high-end electronic products. Most South Asian countries have weak economic foundations, and they mainly import industrial products and capital-intensive products from China. Therefore, China's economic growth has invested more in high-end electronic products and slowly reduced its investment in light industry, which has led to a gradual decrease in China's exports to South Asia and a decrease in trade complementarity. [12]

(2) In terms of population, demographic factors have a significant positive impact on bilateral trade. The $LNPOP_{it}$ coefficient of 15.4658 indicates that for every 1% increase in the population of South Asian countries, China's exports to South Asian countries will increase by 15.4658%, and the population growth of South Asian countries will also accelerate trade exchanges, which is contrary to expectations, the reason may be that the population of trading countries drives the increase of market demand and labor force, which leads to an increase in trade volume. The $LNPOP_{jt}$ coefficient of 0.9567 means that for every 1% increase in China's population, China's exports to South Asia increase by 0.9567%. The total population coefficient of importing countries is much greater than that of exporting countries, indicating that population changes in importing countries have a more prominent impact on trade.

(3) In terms of distance, it has a significant effect on trade and is not in line with expectations. Previous scholars' research distance generally hinders trade, but the result of this paper is positive. The reason may be that with the development of the economy, the

infrastructure of various countries is constantly improving, the effect of distance gradually weakens. It also shows the gradual improvement of transportation facilities in China and South Asian countries to promote trade efficiency.

(4) Boundary factors, and trade have a significant effect at the 1% level. The correlation coefficient is negative. The reason may be that the poor geographical environment between China and South Asian countries hinders imports from trading parties. It is also possible that with the improvement of transportation levels, sea and air transportation will become more convenient, making the existence of common borders less impact on trade , Countries have also tended to develop globally, and cultural and religious factors have gradually weakened.

3.3 Analysis of regression results of trade inefficiency model

This paper adopts a one-step method to perform regression analysis on trade inefficiency items. According to the frontier4.1 regression results, the results are shown in Table3:

Table 3. Estimated results of trade inefficiency model.

SFGM				Trade inefficiency model			
Variable	Coefficient	Standard-Error	T-Ratio	Variable	Coefficient	Standard-Error	T-Ratio
Constant	-86.2828*	49.7318	-1.7350	*Constant*	-1.4778	1.4605	1.0118
$LNGDP_{it}$	0.5981***	0.0663	9.0223	FTA_{ijt}	-2.5898***	0.9788	-2.6459
$LNGDP_{jt}$	0.3159***	0.0752	4.2019	AHS_{jt}	0.1174***	0.0334	3.5156
$LNPOP_{it}$	4.1463*	2.3551	1.7605	GI_{jt}	0.0587***	0.0100	5.8975
$LNPOP_{jt}$	0.7775***	0.0116	66.8541	TF_{jt}	-0.0359**	0.0175	-2.0545
$LNDIS_{ij}$	0.0192	0.2235	-0.0859	FF_{jt}	0.0648*	0.0356	1.8216
BOR_{ij}	-0.1587***	0.0422	-3.7563	WEF_{jt}	-2.2187***	0.2962	-7.4904
				PR_{jt}	0.1684*	0.0978	1.7208
				σ^2	0.6818***	0.0967	7.0496
				γ	0.9898***	0.0028	351.9917
				Log likelihood function	-34.3862		
				LR Test	160.6190		

From the above table, the value of γ is as high as 0.9871, which is very close to 1, which further shows the suitability of choosing a time-varying SFGM. In the trade inefficiency model, the coefficient is positive, indicating that this factor is a promotion factor of the inefficiency model. It is the reason for hindering trade, and the coefficient is negative, indicating that this factor is the hindering factor of the inefficiency model and the reason for promoting trade.

(1) Free Trade Agreement (FTA_{ijt}) is negatively correlated with inefficiency and is significant, which can reduce the hindrance of trade inefficiency factors and improve trade efficiency. Among the inefficiencies, the coefficient of the free trade agreement is the largest, indicating that the signing of the free trade agreement is an important factor in promoting bilateral trade. The signing of a free trade agreement between the two countries can lower tariff barriers and promote China's exports to South Asia. Increase the volume of bilateral trade. Therefore, countries should encourage the signing of free trade agreements. In addition

to India, all South Asian countries have signed free trade agreements with China, which can significantly promote their trade. India should also reduce territorial disputes with China and establish free trade agreements with China to increase the freedom of trade between two sides.

(2) Simple average tax rate (AHS_{jt}) for all products. The tax rate is positively correlated with trade inefficiency and is significant. A 1% reduction in the tax rate will increase bilateral trade by 0.1365%. The lower the tax rate, the lower the amount of tax to be paid, and the higher the trade flow between the two countries.

(3) Government integrity: The coefficient of government integrity (GI_{jt}) is positive, which is positively correlated with trade inefficiency, and is significant at the 1% level, which is contrary to expectations. The government has a high degree of credibility, which reduces trade efficiency. The coefficient is very small and only 0.0679, indicating that the degree of influence is small. The possible reason is that the government's credibility is not necessarily measured in trade, and the main factor influencing the trade between the two countries is not government credibility.

(4) Freedom indicators: trade freedom (TF_{jt}), financial freedom (FF_{jt}), TF and the explained variable are positively correlated and significant, indicating that the open free trade market enhances the trade between the two countries, the trade freedom is high, and the country is open to trade Various restrictions on exchanges and increased activity in market trade have a certain stimulus effect on trade. FF and inefficiency are positively correlated, indicating that open financial markets have a certain level of hindrance to the trade between the two countries, and the variables have a significant relationship. The possible reason is that South Asia and China are both developing countries, compared to the financial system of developed countries. It is not perfect yet. The government's laissez-faire financial market may cause chaos in market financial transactions, and developing countries cannot afford the losses caused by the laissez-faire financial market. The index coefficient of the degree of freedom is the smallest among all inefficiencies, indicating that the main factor affecting inefficiency is not the degree of freedom index. The effect of the degree of freedom index is not obvious when the development situation of various countries is unbalanced. The possible reason is relative to the trade and financial level system of developed countries. South Asia and China are not very complete and their forces are relatively small.

(5) Port infrastructure quality (WEF_{jt}), the coefficient is negative and the 1% significance test is passed. The possible reason for this phenomenon is that the South Asian countries are relatively backward and the port infrastructure is not perfect, so the two countries should strengthen infrastructure construction to promote trade development. China should give full play to the role of the Belt and Road and economic corridors to help South Asian countries build infrastructure to promote the long-term development of bilateral trade.

(6) Political risk (PR_{jt}) PR is positive and significant. The country's political risk is positively correlated with inefficiency, and negatively correlated with China's exports to South Asian countries, which is consistent with expectations. The ethnic and religious contradictions and political institutions in South Asian countries are very severe. Therefore, there are risks in the exchanges between the two countries. The two countries should work together to resist political risks and work together to let the economic corridor play its advantages, so that the people can benefit from the Belt and Road Initiative and reduce political disputes.

3.4 Calculation of the trade potential between China and South Asia

Through the formula: $T_{ijt}=T_{ijt}/T_{ijt}$ Calculate the value of trade export potential, the value of trade efficiency is between 0 and 1, as the value of trade efficiency increases, it means that the country's trade efficiency with my country is higher, and vice versa. Because the BC95 model has human-made inefficiencies and changes with time, this paper uses the one-step method to measure trade efficiency.

3.4.1 Analysis of trade efficiency

The trade efficiency value is sorted according to the frontier regression results to get the annual trade efficiency of South Asian countries, and draw a line chart to see its trend. It can be seen from the data and graphs that, generally speaking, the trade efficiency between China and South Asian countries shows a fluctuating upward trend and then declines, but the trend is not very obvious, which also shows that China and South Asian countries have great potential for trade development. 2000-2008 is generally on the rise. In 2008, the financial crisis encountered a decline in trade efficiency and continued to rise. In 2014, China initiated the Belt and Road Initiative and signed free trade agreements with various countries to help South Asian countries promote infrastructure and reduce bilateral tariff rates. The open trade market has greatly facilitated trade flows. In recent years, trade efficiency has shown a downward trend. There have been some political disputes in various countries, the normal implementation of economic corridors, and the interference of other countries have reduced trade efficiency. In terms of specific countries, each country has different levels. Pakistan and Sri Lanka have maintained good trade relations with China, indicating that the two countries have maintained close trade relations. The China-Pakistan Economic Corridor has also played a significant role. The trade between the Maldives and China has shown an obvious upward trend year by year. In 2017, the Maldives signed a free trade agreement with China, and the trade volume has increased significantly. The efficiency of Sino-Indian trade continues to rise, but in recent years, political border conflicts between China and India and internal risk issues within the country, as well as India's control of South Asian countries to prevent other South Asian countries from cooperating with China, have reduced the cooperation between the two countries. The trade relationship between Nepal and China shows a fluctuating trend. Due to the natural geographical environment and China separated by the Himalayas, hindering the trade and transportation between the two countries and India's suppression of Nepal, the trade relationship between China and Nepal is not very close. In the context of the Belt and Road Initiative from 2010 to 2014, China-Nepal trade efficiency achieved leapfrog development, and then declined. The development of Bhutan and China has been slow. For two decades, there has been no obvious trend of trade development. The two countries have not established diplomatic relations. On the one hand, Bhutan has always adopted a closed attitude and its economic development is backward. What's more, Bhutan's trade has long been dependent on India, and India has also restricted Bhutan. With economic development, Bhutan also hopes to get rid of its dependence on India, but now Bhutan is still adopting a wait-and-see strategy for the One Belt One Road strategy. China's exports to Bhutan are mostly agricultural products and low-end products, and Bhutan's imports are also low-end products. Therefore, the structure of commodity trade between the two countries has not changed. With the passage of time and the development of bilateral relations, the two countries are expected to establish a friendly relationship of mutual assistance and mutual assistance under the "Belt and Road" initiative.

Table 4. China's trade efficiency with South Asian countries.

Year	Bangladesh	Bhutan	Sri Lanka	Maldives	Nepal	Pakistan	India
2000	0.51	0.07	0.89	0.05	0.59	0.37	0.20
2001	0.50	0.05	0.73	0.07	0.40	0.42	0.22
2002	0.85	0.03	0.93	0.14	0.44	0.93	0.47
2003	0.56	0.05	0.75	0.08	0.26	0.75	0.30
2004	0.68	0.01	0.88	0.16	0.30	0.84	0.44
2005	0.76	0.01	0.95	0.31	0.30	0.94	0.57
2006	0.83	0.00	0.94	0.22	0.34	0.94	0.77
2007	0.72	0.06	0.94	0.28	0.41	0.96	0.93
2008	0.79	0.08	0.91	0.28	0.32	0.92	0.96
2009	0.69	0.04	0.82	0.32	0.31	0.82	0.88
2010	0.89	0.01	0.84	0.43	0.46	0.88	0.94
2011	0.86	0.10	0.95	0.54	0.61	0.87	0.95
2012	0.79	0.08	0.91	0.38	0.90	0.86	0.89
2013	0.85	0.08	0.91	0.42	0.91	0.90	0.84
2014	0.89	0.05	0.91	0.40	0.88	0.94	0.83
2015	0.93	0.03	0.94	0.59	0.30	0.96	0.85
2016	0.91	0.02	0.92	0.94	0.30	0.96	0.80
2017	0.87	0.02	0.84	0.85	0.30	0.95	0.83
2018	0.89	0.04	0.80	0.93	0.29	0.88	0.83
2019	0.82	0.03	0.75	0.81	0.37	0.84	0.77

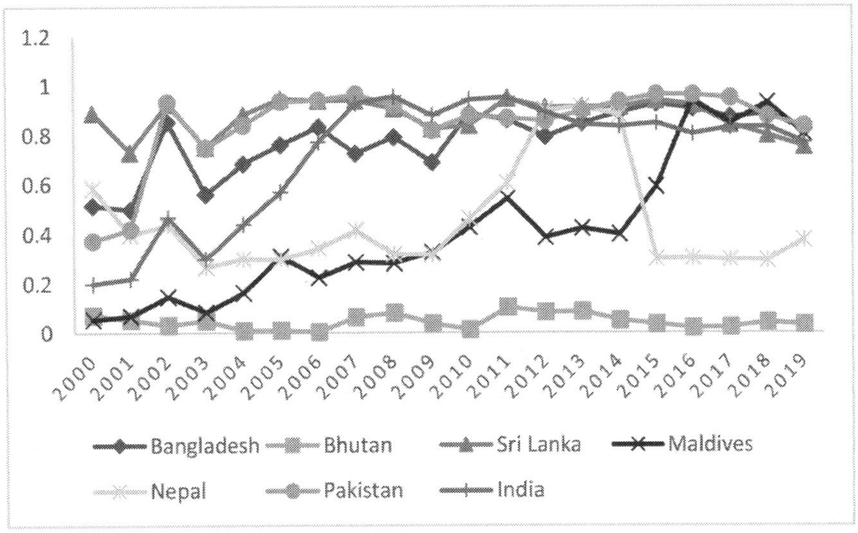

Fig. 1. China's trade efficiency trend with South Asia.

In order to better understand the trade potential of China and South Asian countries, according to the regression results, the average trade efficiency, average real export value, average trade potential value and expansion space of each country from 2000 to 2019 are made to better compare and analyze the trade development level of different countries. The results are shown in Table 4. As can be clearly seen from the table rankings, the average trade potential ranking is followed by Sri Lanka, Pakistan, Bangladesh, India, Nepal, Maldives and Bhutan. Since the friendly exchanges between China and Pakistan and Sri Lanka in 2014, the volume of trade has expanded rapidly. It has surpassed India to become the first and second trade potential partner of China in South Asia. India's trade is declining because of its recent political disputes with China. Pakistan's trade efficiency value with China in 2019 was 0.84, making it the most efficient country in the trade with China. This shows that bilateral trade cooperation has made remarkable progress. During the epidemic, the two countries also helped each other with epidemic prevention materials, and the relationship between the two countries has heated up after the epidemic. The trade efficiency of Nepal and Maldives with China is low at around 0.4 and there is still a lot of room for development. But Bhutan's trade efficiency is very low, only 0.04, so there is a lot of room for development. Therefore, China and Bhutan should seize the opportunity to remove the obstacles to bilateral exchanges, form a long-term trade plan, establish diplomatic relations with each other, and join the One Belt and One Road.

Table 5. Average values of indicators between China and South Asian countries from 2000 to 2019.

Country	eff.-est.	Rank	Actual export volume (million)	Export potential	Expand space
Sri Lanka	0.88	1	3511.9	4010.75	0.14
Pakistan	0.85	2	830.46	982.03	0.18
Bangladesh	0.78	3	732.55	939.83	0.28
India	0.71	4	226.27	317.2	0.4
Nepal	0.45	5	79.78	177.64	1.23
Maldives	0.41	6	10.6	25.81	1.44
Bhutan	0.04	7	0.66	15.1	22.03

Data source: Calculated based on the results of the trade inefficiency model.

Note: Expansion space = (export potential / actual export value-1) * 100%

4 Conclusions and policy recommendations

4.1 Basic conclusion

This paper uses the time-varying stochastic frontier gravity model to select data from the past 20 years to study the relationship between China and South Asian countries' trade, introduces trade efficiency and artificial inefficiency factors to measure trade efficiency, analyzes related factors and compares the results with expectations to draw the corresponding conclusion.

Judging from the results of the SFGM analysis, the main factors affecting the trade between China and South Asian countries are the per capita GDP of the two countries, the total population, and whether there is a common boundary. The result of geographic distance is not significant. It is concluded that the growth of the economic scale and population of the two countries has an obvious promotion effect on bilateral trade, while the growth of China's GDP has an obstructive effect on bilateral trade. The distance between the two countries in trade is to enhance the utility of bilateral trade, and may be the weakening of the distance utility. The existence of a common border between the two countries can also effectively promote bilateral trade. The non-efficiency model has great obstacles to bilateral trade. Among them, whether to sign a free trade agreement, the degree of trade freedom, and the quality of port infrastructure play a role in promoting trade and are all significant. The simple average tax rate of all products, government integrity, financial freedom, and political risks hinder trade and are all significant.

From the perspective of trade efficiency, China's trade efficiency value with Bangladesh, Pakistan, and Maldives is high in 2019, and it has certain development potential with India, Nepal, and Sri Lanka. It has the lowest average trade efficiency with Bhutan. China's trade has huge potential for development.

4.2 Countermeasures and suggestions

(1) China should establish friendly diplomatic relations with South Asian countries and play the role of the Belt and Road and the Maritime Silk Road. In the context of the COVID-19, the development of new industries has been continuously promoted. For example, the operation of Internet platforms, home office, biomedicine, online teaching, etc. will continue to develop, 5g technology will develop against the trend due to the epidemic, and the pace of the information age will accelerate, while the economic and technological level of South Asian countries is backward, and external technical equipment and materials are more needed Therefore, China should join hands with South Asian countries to overcome difficulties. China should also speed up the cross-border e-commerce transaction mechanism with South Asia, reduce taxes and fees, and stabilize industrial supply to expand trade volume. [13] Bhutan has not established diplomatic relations with China so far, indicating that the two sides still have concerns. Bhutan has also been extremely backward in economic development due to a long-term closed attitude to the outside world and has not yet opened up ideologically. However, India's suppression in recent years has led to some signs that Butan is getting rid of this situation, and the two sides should seize the opportunity of the Belt and Road to enhance mutual trade. Hoping that Bhutan's trade with China will improve in the near future. India has not participated so far, China and India should also remove factors that hinder the exchanges between the two countries and join the Belt and Road Initiative. China and South Asia should also promote the construction of a free trade zone and further deepen open cooperation. If China and South Asian countries with high trade efficiency continue to maintain friendly trade exchanges, the trade volume between the two countries will be further deepened to develop new goods for trade exchanges. For countries with extremely low trade efficiency, the two countries should lower trade barriers and adopt a positive attitude to promote trade development.

(2) Promote the financing function, and the government strengthens the management of the financial market. Project construction in China and South Asian countries generally has the characteristics of long cycle and high capital demand. The Asian Infrastructure Investment Bank and the Silk Road Fund should also play their financing role to help South Asian countries provide financial support. To constantly improve the financial system services, and continue to promote the internationalization of the RMB. Continue to promote the construction of infrastructure. With the continuous expansion and improvement of the "Belt and Road", China has helped many countries in South Asia to carry out infrastructure construction, but there are still many projects that have not been completed. China should use its advantages in infrastructure such as roads and railways to promote the development of China and South Asia.

(3) According to the situation of different countries, implement a differentiated trade model that meets the needs of the two countries. [14] The economic situation, political situation and trade environment of South Asian countries are different. The Chinese government should coordinate the various demands of different countries to strengthen the

understanding of the culture, politics, economy and other aspects of South Asian countries and establish sustainable trade relations with South Asia. China has always been in a surplus in exports with South Asia, and trade exchanges account for a small proportion of China's exports. This reflects the trade issues between China and South Asia and also shows that there is a lot of room for expansion in trade cooperation. [15] The trade between China and South Asian countries is mostly capital-intensive products, and China's exports are shifting towards high-end electronic products. The two sides should expand new industries for trade exchanges, and continuously improve and innovate export products to promote trade transformation between the two sides. Countries should unite to resolve domestic and international security issues. South Asia has always been threatened by terrorism. The Global Terrorism Index shows that South Asia is most affected by terrorism in the world. South Asian countries should improve national management and control capabilities, reduce inequality in society, and take measures to ease people's sentiments to reduce terrorism.

(4) Strengthen cooperation and exchanges between the western region and South Asia. China's Tibet and Xinjiang are connected with South Asian countries. They should give full play to their geographic advantages and extend their territory to South Asia with the west as its territory. Therefore, China should strengthen assistance to the west by establishing industrial parks, strengthening infrastructure construction in the west, and providing high-tech equipment. It not only promotes the development of China and South Asia, but also gives a lateral impetus to the strategy of developing the western region.

References

1. A.Islam, R.Salim, & H.Bloch, Does Regional Integration Affect Efficiency and Productivity Growth? Empirical Evidence from South Asia. Review of Urban and Regional Development Studies, **28(2)**, 107–122. (2016)
2. R. Kumar, South Asia: Multilateral trade agreements and untapped regional trade integration. Int J Fin Econ. 2020; 1– **13**. (2020)
3. N. Gul, & H.M Yasin, The Trade Potential of Pakistan: An Application of the Gravity Model. Lahore Journal of Economics, **16**(1), 23–62. (2011)
4. N.Akhter, E.Ghani, Regional Integration in South Asia: An Analysis of Trade Flows Using the Gravity Model. Pakistan Development Review. 2010;**49**(2):105-118. (2010)
5. Y Y Zong. Analysis on the growth potential of agricultural trade between China and South Asian countries[D]. Shandong University of Finance and Economics (2018).
6. Z X Liu, H H Chen. An Empirical Study on Energy Intra-Industry Trade between China and ASEAN[J]. Coal Economic Research, 2019, **39**(09): 56-62. (2019)
7. Y J Cheng, C H Dong. Research on the efficiency and potential of trade between China and South Asian countries from the perspective of "One Belt and One Road"——An Empirical Test Based on the Heterogeneous Stochastic Frontier Gravity Model[J]. Journal of Lanzhou University of Finance and Economics, 2020, **36**(02) :94-105. (2020)
8. J Liu. Research on the trade potential and influencing factors between China and Central Asia and South Asia [D]. Guangdong University of Foreign Studies, (2020).
9. L Wang, S Chen. Research on the Trade Efficiency and Potential of Bangladesh-China-India-Myanmar Economic Corridor——Based on the Analysis of Stochastic Frontier Gravity Model[J]. Asia-Pacific Economics, (2019)

10. Z G Gao, F Tian. Research on the impact of trade facilitation level on China-Pakistan export trade efficiency under the background of China-Pakistan Economic Corridor[J]. South Asian Studies, (2019)
11. L Li, S C Ma. An Empirical Study on Trade Competitiveness and Complementarity between China and South Asia[J]. Price Monthly, (04): 49-58, (2020)
12. Y X Li. The cooperation basis, promotion opportunity and realization mechanism of China and South Asia trade under the "One Belt and One Road" strategy. Gansu Social Sciences, (2016)
13. L H Xu. China's New International Competitive Advantages of Foreign Trade in South Asia and Southeast Asia[J]. South Asia and Southeast Asia Studies, (2020).
14. J Wei, H Y Zhang. Potential analysis of import trade between China and Central and Eastern European countries: an empirical study based on the time-varying stochastic frontier gravity model [J]. Regional Economic Review, (2020)
15. Y M Wu. Research on China-South Asia trade cooperation under the background of "One Belt and One Road" [D]. Yunnan University of Finance and Economics, (2016)

Research on Performance Evaluation of Diversified Government Quality Work and Analysis on Application Practice in Nei Mongol

Lei Feng, *Juxiu* Huang* *and Jingxing* Liao

China National Institute of standardization, Beijing 100191, China

Abstract. The evaluation of public satisfaction with government quality work is an evaluation form to evaluate government performance from the perspective of the public. The evaluation process is open and transparent, and the results are relatively objective and fair. Taking the application practice in Nei Mongol as an example, in this paper, an index framework is designed and constructed, 12 leagues and cities in the whole region are covered by the investigation, and the actual effect of local quality work is explored and analyzed in combination with big data technology so as to provide enlightenment and reference for relevant research work in the quality field.
Keywords: Quality work, Satisfaction, Application Practice.

1 Introduction

The social public satisfaction evaluation of government quality work is one of the effective methods to connect quality management departments and the public [1]. The public is an external evaluation subject introduced by the government departments when extending the evaluation angle to the outside on the basis of internal evaluation. On the one hand, the introduction of the external evaluation subject promotes the transparency and openness of quality performance evaluation; On the other hand, it promotes the diversification of evaluation paths and evaluation methods, providing a new idea for accelerating the gradually scientific and standardized government performance evaluation [2]. Taking the survey data of Nei Mongol government's quality work satisfaction as an example, based on the statistical survey results, in this paper, the actual effect of local quality work was explored by multi-dimensional research of the evaluation results such as the overall analysis, regional analysis, key area analysis, influencing factor analysis and others, providing enlightenment and reference for the research of quality work performance evaluation.

* Corresponding author: 1053038950@qq.com

2 Analysis on application practice in Nei Mongol

The evaluation of social public satisfaction with the quality work of Nei Mongol government adopted Likert 5-level scaling method [3]. The evaluation content includes five major fields of product quality, engineering quality, service quality, environmental quality and quality awareness, with a total of 5 first-level indicators, 13 second-level indicators and 31 third-level indicators. The evaluation covered 12 leagues and cities in the autonomous region. 3813 valid samples were collected through the questionnaire survey of residents' satisfaction. The SPSS reliability analysis, KMO value and Bart spherical test results showed that the KMO value was 0.973, the significance was less than 0.05, the reliability value was 0.943, and the reliability and validity were good.

2.1 General situation

The Nei Mongol Autonomous Region Government scored 82.65 points in the social public satisfaction evaluation of quality work. Therein, the product quality satisfaction was the highest, with 83.83 points; the second was quality consciousness, with 82.74 points; the satisfaction scores of engineering quality, service quality and environmental quality were 82.33, 82.25 and 82.10 respectively. Please see the following figure 1.

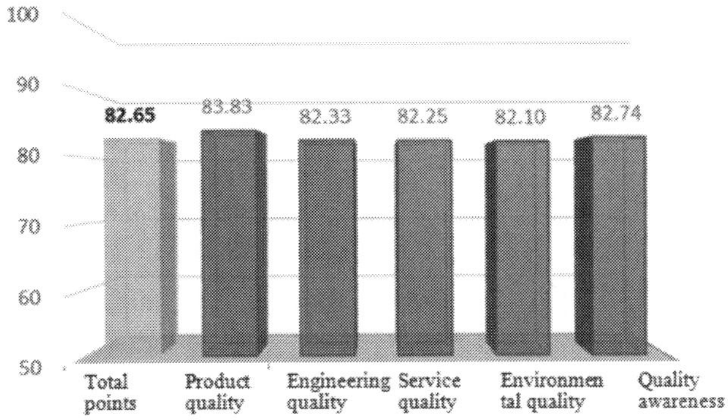

Fig. 1. Overall situation of social public satisfaction evaluation of Nei Mongol government quality work.

2.2 Regional analysis

The evaluation results of regional quality work satisfaction showed a trend of being "high in the middle and low in the west". The satisfaction in the central region was the highest, with 83.53 points, which was 0.88 point higher than the overall level of the autonomous region. The satisfaction in the eastern region took the second place, with 82.84 points. The western region was relatively low, with 79.8 points. Among the 12 leagues and cities, 7 leagues and cities' satisfaction were higher than the average level of the whole region, among which Tongliao, Wulanchabu, Erdos, Baotou and Hohhot were in the top 5 with scores of 83.91, 83.89, 83.51, 83.48 and 83.24. The satisfaction of 5 leagues and cities was lower than the average level of the whole region, with Xilingol, Wuhai and Alxa at the bottom 3, scored 81.77, 79.56 and 77.53 respectively. The imbalance of quality development among regions still existed in the whole region, and the quality advantages in some regions were not prominent enough. Please see the figure 2.

 https://doi.org/10.1051/shsconf/20219601011

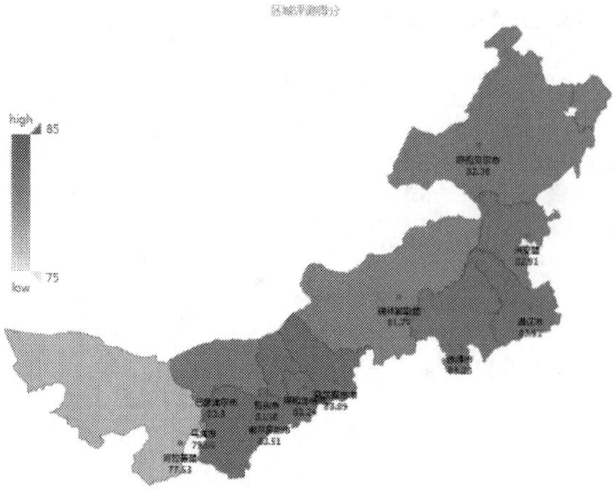

Fig.2. Regional Evaluation Results.

2.3 Evaluation situation of key areas

In the evaluation of product quality satisfaction, the drug quality satisfaction was higher while the special equipment quality satisfaction was lower. Judging from the monitoring indicators, the satisfaction with drug quality was the highest, with 85.10 points. The satisfaction with the quality of special equipment was relatively low, with 82.45 points. The quality satisfaction of agricultural products, household durable goods, food and daily consumer goods was 84.76 points, 84.05 points, 83.96 points and 83.12 points respectively. The main influencing factors that led to the low satisfaction evaluation score of the "special equipment quality" were analyzed as follows: 32% of the respondents thought that there was a lack of regular maintenance, 24% thought that there were potential safety hazards, and 24% thought that maintenance personnel were not in place.

In the evaluation of engineering quality satisfaction, the traffic engineering quality satisfaction was slightly higher than the constructional engineering quality satisfaction. Judging from the monitoring indicators, the satisfaction with traffic engineering quality was relatively high, with 82.70 points. While, the quality satisfaction of the constructional engineering was relatively low, with 81.97 points. Judging from the main factors that affect the evaluation of engineering quality satisfaction, the poor sound and heat insulation effect of the houses and the cracks appeared on the wall surface and floor slabs of the houses were the main reasons that affected the public's evaluation of the quality of the constructional engineering. Unreasonable design and construction, jerry-building and poor quality of main structures were the main reasons that affected the public's evaluation of traffic engineering quality [4].

In the evaluation of service quality satisfaction, the life service satisfaction was low. Judging from the monitoring indicators, the quality satisfaction of the productive services such as bank and long-distance and short-distance transportation services was high, with 83.49 points, 83.07 points and 83.01 points respectively. The satisfaction with the living services provided directly to residents, such as providing for the aged and family decoration, was relatively low, with 79.98 points and 80.7 points respectively. Among many influencing factors, insufficient nursing facilities, low quality of nursing staff and shortage of services were the main factors affecting the quality of pension services. Poor market order, poor personnel quality and lack of service supervision were the main factors affecting the quality of family decoration services.

There was still a big gap between the evaluation results of the environmental quality satisfaction and the expectations of the public. Judging from the monitoring indicators, the public's satisfaction with the water environment quality was still at a relatively low level, with a score of 81.55 points, which was the lowest in the 13 second-level indicators. The reasons were as follows: 40% of the unsatisfied public thought that domestic waste pollution was serious, 27% of the dissatisfied public thought that the industrial emissions were too high, and 17% of the unsatisfied public thought that the agricultural pollution was serious. The satisfaction with air quality was 82.65 points, on which industrial exhaust gas, automobile exhaust gas and domestic exhaust gas pollution had a certain negative impact.

Quality publicity and information disclosure were still needed to be further strengthened in the evaluation of quality awareness satisfaction. Local residents paid high attention to quality, with 84.53 points. Residents fully affirmed the improvement of local quality work and the handling of quality complaints, while saying that quality publicity and information disclosure work are to be further strengthened. According to further investigation, 37% of the residents interviewed who were "not very satisfied" or "very unsatisfied" thought that the disclosure of quality and safety information was not timely or accurate, 35% of the residents thought that the open and publicity channels of quality information were few, and 26% of the residents thought that the publicity activities were few and small in scale.

2.4 Analysis of influencing factors

According to the monitoring results, the satisfaction of different gender respondents with government quality work varied slightly, and the scores were tended to be consistent, with men scoring 82.81 points and women scoring 82.43 points. From the point of age, the youth group showed the highest satisfaction with the government's quality work, with 82.95 points. The middle-aged group took the second place, with 81.18 points. The satisfaction of the elderly group was relatively low, with 79.48 points. From the perspective of academic qualifications, the respondents with high academic qualifications were more satisfied with the quality work of the government. Among them, the respondents with bachelor's degrees showed the highest satisfaction of 83.47 points, while the satisfaction degree of the respondents with primary school qualification or below was 79.65 points. Judging from different household registration types and income situations, the satisfaction with the government quality work of the rural respondents was lower than that of the urban respondents [5]. Among the respondents who revealed the average monthly income of their families, the satisfaction with the quality work increased with the increase of their income, with the highest satisfaction degree of 84.71 points from the respondents whose average monthly income of urban families was RMB 10001-20000 yuan.

3 Conclusion

Based on the actual measurement and analysis of Nei Mongol government quality work satisfaction through the diversified government quality work performance evaluation research, and combined with the measurement results, the following suggestions are put forward for Nei Mongol to carry out quality improvement actions, strengthen overall quality supervision, comprehensively improve quality level, and promote the implementation of the construction of province of high quality.

3.1 Focusing on the quality needs of the public so as to benefit people's livelihood by taking well-targeted steps

Safeguarding and improving people's livelihood is the fundamental task of social construction, and the fundamental way to improve the satisfaction evaluation results of government quality work. First of all, in order to make up for the shortage of people's livelihood development, governments at all levels in Nei Mongol Autonomous Region should focus on the "demands" and "concerns" of the masses in product quality, engineering quality, service quality and environmental quality, and strengthen the supervision of pension services, family decoration services, tourism services, water environment quality and other fields. Secondly, governments should strengthen quality social supervision and public opinion supervision, establish and improve the feedback mechanism of quality information transmission, strengthen comprehensive quality management, protect people's quality rights and interests, and enhance residents' senses of acquisition and happiness.

3.2 Learning from the preponderant experience of quality to promote coordinated regional development

Facing the current situation of unbalanced quality construction level in various leagues and cities, we should thoroughly implement all-round development and balanced development, and give full play to the preponderant experience of high-level regions in quality construction through the endogenous power development mode of economic development to drive the quality construction development in backward regions.[6] Adhering to carry forward the advantages and characteristics of various regions, we should, through exchanges, experience and resource sharing and other ways and means, achieve the quality improvement work of "making the advanced better, driving the development of backward by the advanced and realizing mutual development", improve the public's satisfaction with the quality work, and effectively achieve the coordinated development of regional quality work to enable the whole people to share the results of quality development.

3.3 Consolidating quality infrastructure and promoting high-quality development

The quality foundation is an important guarantee to carry out quality improvement actions and to promote high-quality development. [7]The evaluation results showed that the quality development of products, engineering, services and other fields in the whole region still lagged behind the needs of economic and social development and the people's needs for a better life, which is related to the low level of quality infrastructure development. It is suggested to further strengthen the construction of quality infrastructure in the whole region. According to the regional characteristics and the advantages of traditional industries, we should promote the "internet plus NQI quality model", strengthen the technology chain design and integrated application of "measurement + standards + inspection and testing + certification and accreditation" technology, strengthen the overall integration and construction of quality infrastructure, and effectively improve the quality foundation support capacity so as to serve the high-quality development of the economy.[8]

Acknowledgements

This paper is funded by the Government Quality Monitoring, Evaluation and Assessment System Improvement and Implementation Project (582019C-7110) and the Quality Infrastructure Collaborative Service Application Research Project (582019C-6930).

References

1. M L Morris, R Tripp, and A A Dankyi. Adoption and Impacts of Improved Maize Production Technology: A Case Study of the Ghana Grains Development Project, International Maize and Wheat Improvement Center (CIMMYT) (1999)
2. J O Ouma, F M Murithi, W Mwangi, H Verkuijl, M Gethi, and H D Groote. Adoption of Maize Seed and Fertilizer Technologies in Embu District, Kenya, International Maize and Wheat Improvement Center (CIMMYT) (2002)
3. X G Luo and H Y Zhang, Y Hong. Evaluation dimension analysis of government service quality SERVQUAL [J]. Administrative Forum, **3**,35-37 (2008)
4. G F Cui. On the Development of Government Service Quality Management [J]. Hubei Social Sciences, **8**,37-38(2005)
5. W B Li and L H Lai, Government Performance Satisfaction and Residents' Well-being: An Empirical Study in Guangdong Province [J]. China Administration, **8**,53-57(2013)

SHS Web of Conferences **96**, 01012 (2021) https://doi.org/10.1051/shsconf/20219601012
IAFSM 2020

The Construction and Characteristics of Foreign Army's Electronic Countermeasure Equipment

Cuodong Zhang*

Naval Research Academy, Beijing 100161, China

Abstract. Based on the introduction of the development of electronic countermeasures, this paper introduces the composition, construction and application characteristics of foreign electronic countermeasures equipment, and puts forward some Suggestions on the development of electronic countermeasures.
Keywords: Foreign army, Electronic countermeasure equipment, Development of electronic countermeasures

1 Introduction

Electronic countermeasure, also known as foreign electronic, electronic warfare, the beginning is mainly used for launch on radio signals or using radio direction finding itself deceive signal is disturbed, practical use for the first time was in 1904, the Russian naval warfare, Russian shore-based radio and interfere with the Japanese success before a picket ship shelling guide communication, make it failed to reach a sneak attack.

The Vietnam War and the Middle East War developed the theory of systematic electronic countermeasures and formed the combat doctrine, and electronic countermeasures developed from "guarantee means" to an important "combat means" capable of carrying out offensive operations and dealing with precision-guided weapons.

The Gulf War is a significant sign of the development of system confrontation, and the application of electronic confrontation has reached an unprecedented scale and level, laying the battle idea of "taking electronic confrontation as the guide and running through the war" in modern war.

In the subsequent local wars, electronic confrontation played a more and more significant role, and its development attracted the attention of all countries. By early 2017, the US Department of Defense has released its first ELECTRONIC Warfare Strategy, which describes how the US Department of Defense will conduct electronic warfare and electromagnetic spectrum operations in a wider range. It not only provides strategic guidance for the US electronic warfare development, but more importantly, further promotes the establishment of electromagnetic spectrum independent operational areas.

* Corresponding author: 380055262@qq.com

2 System composition and characteristics

In view of the most perfect and mature electronic countermeasures construction and actual combat application of the US military, which represents the highest development level of foreign countries, this paper takes the US military as an example to analyze the construction and application of foreign electronic countermeasures.

In March 1993, the department of defense staff, chairman of the joint meeting with memo (CJCS) form is issued policy documents, determine the electronic warfare are defined as the use of electromagnetic energy and orientation to control electromagnetic spectrum or attack any enemy military action, including electronic support (ES), electronic attack (EA) and electronic protection (EP) three main parts, according to the actual equipment system structure and application situation, should add operations support related parts.

2.1 Electronic support

Electronic reconnaissance has the advantages of long distance, wide coverage and strong identification ability. It can "conceal" the electromagnetic situation of naval battle field, collect electronic intelligence, and analyze the force dynamics and activity rules. It is the premise and basis for making operational planning and implementing effective interference. American military experts, when analyzing the reasons for winning the Gulf War, believe that "the most lethal weapon in the Gulf War is not missiles and fighter planes, nor warships and tanks, but the huge electronic reconnaissance and intelligence system deployed in the region". In the war, the americans the use of a large number of space-based power electronic reconnaissance, air and ground, including the invaders, magnum and so on more than 20 electronic reconnaissance satellites, eight EC - 130 - h, 28 EP - 3 e, 18 RC - 12, 9 RC - 135, 15 U - 2 and so on electronic reconnaissance aircraft, 39 unmanned reconnaissance aircraft, etc., to build a seamless reconnaissance monitoring network, the real-time target detection, identification, interference by GPS, combat assessment plays an important role in such aspects.The electronic reconnaissance construction of the US army has two remarkable characteristics: "complete means" and "excellent intelligence" :

First, comprehensive means - means construction, pay attention to system application. With electronic reconnaissance satellites, electronic reconnaissance aircraft, unmanned aerial vehicles and reconnaissance ships as the core, the US military has built a complete reconnaissance, surveillance and intelligence collection system, which is capable of normalized reconnaissance, surveillance and continuous tracking of hot spots and sensitive events. If electronic reconnaissance satellites are vigorously developed, there are about 30 fourth-generation and some third-generation electronic reconnaissance satellites in service, which can be divided into low-orbit Marine surveillance satellites and synchronous orbit electronic reconnaissance satellites. The low-orbit ocean surveillance satellites, mainly "Baiyun" and its successor "space-based Wide area Surveillance system", cover the frequency band of 154MHz ~ 10GHz. The multi-satellite network can locate the ships on the surface of the sea. The synchronous orbit electronic reconnaissance satellite is mainly "Magnum" and its subsequent "consultant". It covers the frequency band of 100MHz ~ 20GHz and adopts a large umbrella-like phased array reconnaissance antenna with a diameter of about 35 meters. The above satellites are mainly used to detect and receive ground radar, communications, missile telemetry and other signals, which have a great impact on electromagnetic security in our equipment tests, training exercises and other activities.

Second, the intelligence is fine - master fingerprint characteristics, to achieve individual identification. It pays attention to the acquisition of detailed intelligence, which is typical to collect fingerprint characteristics of radiation source, complete target individual identification, and be able to distinguish target platform number (such as ship's hull number) and individual. In the 1970s, the NAVAL Research Laboratory began to develop the radiation source individual identification Equipment (SEI). With the continuous funding and project support, the individual identification database was gradually integrated into the coastal, sea, air, sky and submarine electronic warfare equipment, and the individual identification database was subsequently established and continuously enriched to form the networked application capability. In 2002, the US military organized the "Millennium Challenge 2002" exercise for special verification.Before the exercise, SEI devices collected signal "fingerprint" features to build individual identification database; In the exercise, SEI devices from different platforms on shore, sea and air will transmit the detection and collection results to the command post, and integrate and identify individuals with the support of the database to form a unified and Shared electromagnetic situation and target situation, so as to realize cross-regional and continuous tracking and monitoring of the designated platform.

In terms of application, it pays attention to regular intelligence gathering, perceiving battlefield situation in wartime, detecting threat targets and guiding the implementation of interference or anti-radiation strike. In peacetime, reconnaissance finds unknown radiation sources, and deeply analyzes the acquisition of strategic and tactical information such as equipment performance, position and deployment, activity rules, usage characteristics, equipment system and development trend.

2.2 Electronic attack

When carrying out air penetration and attack missions, the US Navy aviation forces have formed and followed a basic principle: "No Fly without EA-6B electronic warfare aircraft";In the Afghan war, the US Army chief of staff also declared that "the army will not fight in the war without 'compass call' (i.e. Ec-130h communication jamming aircraft)", which shows the role of electronic offensive force in the US military combat system.Usually in the war in advance, throughout the whole mission, such as the Gulf War air raid 15 minutes before, the US army carried out the "Snow" electronic warfare operation, sent EF-111A, EA-6B, EC-130H electronic warfare aircraft to carry out a comprehensive "electronic bombing" on the Iraqi deep radar network, communications system, cut off the Iraqi army's electronic "nerve", electronic "eyes and ears";After the air attack began, f-4G anti-radiation missile was sent to attack the aircraft companion and launch "Ham" anti-radiation missile for hard destruction while suppressing the interference of the Iraqi army's warning, detection and command and communication system.

Therefore, the US military attaches great importance to the development of electronic offensive equipment, and has built a complete equipment system including electronic warfare aircraft, a series of jamming pods, anti-radiation weapons and air launch bait. The main features are as follows:

The first is the coordinated development of soft interference and hard destruction. Heavily in the construction of EA - 18 g electronic warfare aircraft, the next generation of interference, pod (NGJ), and other soft interference power at the same time, enhance the anti-radiation strike capability, with the Italian air force joint AGM - 88 e advanced anti-radiation missile, improve the passive seeker, extension, increase of active millimeter wave terminal guidance radar reconnaissance distance, improve precision and ability to resist shutdown, resistance to movement.

Second, systematic development and application. By means of electronic warfare aircraft, air-launched decoy and missile-borne jammer, the system of force firepower penetration support jamming is constructed to form the whole combat process countermeasure capability combining the long-distance support jamming, the support jamming with the team, the close screen jamming and the platform self-defense jamming.

Third, both peacetime and wartime. In normal times, there is a clear trend of normal application. Powerful electronic jamming capability can be used to interfere with the target of the other side's offshore and shore-based radiation sources, thus forming electromagnetic deterrence. For example, the south China Sea has repeatedly interfered with the radar of our observation stations and naval fleets.

2.3 Electronic defense

In response to the incoming threats, protect the security of the platform, the U.S. ships, aircraft and other platforms are equipped with perfect defensive electronic countermeasures, the battlefield threat warning, active passive comprehensive interference ability, such as surface ships equipped with SLQ - 32 electronic warfare system, passive SRBOC jamming equipment, outboard active decoy, card, part of the ship equipped with MK59 inflatable Angle reflector. The main features of electronic defense equipment construction of the US army are as follows:

First, to keep abreast of threats and continue to upgrade with advanced technologies. The typical is the SLQ-32 electronic warfare system. The system is equipped with nearly 200 sets of aircraft carriers, cruisers, destroyers and amphibious assault ships.To improve its ability to respond to new threats, the U.S. Military has launched the Spiral Upgrade Program (SEWIP), which "integrates advanced technologies and capabilities in an incremental manner, and on this basis achieves batch upgrades. "Upgrade is completed in two stages, including integrated emitter individual identification ability, increase the bait, card control function, improve the complex electromagnetic adaptability and direction finding accuracy, etc., has improved in the third stage, focus on developing jammer phased array system, improve the performance of active jamming, the defensive electronic countermeasures system has the capability of regional support interference.

Second, comprehensive and comprehensive use of means. In addition to the traditional inboard active jamming and outboard passive jamming, the surface ships are generally equipped with the "Nalka" outboard active decoy jointly developed by the United States and Australia, which USES vector rocket motors to achieve hovering and implement outboard active jamming against incoming missiles. In addition, according to the data, the americans in response to my new anti-ship missiles, emergency start electronic reconnaissance project sensitivity, expanding threat target's found that the low intercept system, at the same time, o burke class destroyer, emergency procurement with MK59 Mod0 inflatable Angle reflector, using 60 surface structure, to simulate ship false targets;Moreover development "warship extended range electronic warfare decoy", is a kind of quick launch unmanned aerial vehicle (uav) radar active decoy system, the system by the folding wings, booster rockets, electric drive system and electronic warfare equipment, equipped with electronic warfare equipment can be a long time, no crack and flexible response to incoming anti-ship missiles to implement electronic countermeasures.

2.4 Combat support

During the Vietnam War, the American A-6A attack aircraft alarm equipment was so deficient that the commander of the aircraft Wing said, "The wing is not fit to operate again until the problem is solved."To solve this problem, the Naval Air Command immediately organized experts to carry out intelligence data analysis, adjust the warning strategy, and verify the actual test flight in China Lake range, and finally the false alarm rate was reduced by 60%.Since then, with the emergence of programmable electronic warfare equipment battlefield electromagnetic environment and increasingly complex, the U.S. military started to start preparing and interference strategy used to combat data for validation of electronic warfare combat support power construction, set up professional research department, to develop the corresponding equipment, the role of increasingly significant, so that the kosovo war, then Pacific fleet electronic attack forces commander John collier said, "no EA - 6 b, in kosovo, NATO is not the plane take off;The EA-6B will never take off without Crane.According to the characteristics of different types of ELECTRONIC warfare equipment, each type of ELECTRONIC warfare equipment of the U.S. Navy has corresponding tactical system support center, among which Crane branch is the ELECTRONIC warfare system support center of EA-6B, EA-18G and SLQ-32.The division has more than 4,000 active, off-duty civilian and contract employees, equipped with simulation LABS, hardware-in-the-loop simulation centers, and outfield verification firing ranges.

The MAIN functions of the US Army electronic Warfare Support system include target analysis and modeling, combat data generation and verification, jamming strategy optimization, electronic warfare mission planning, replay assessment and simulation training. The specific procedures are as follows:

One is to use the acquired battlefield intelligence to accurately grasp and evaluate the threat target and generate the target database.

The second is the programming and verification of combat data. According to the published electronic warfare target data, all electronic warfare equipment are simulated and tested to determine the effectiveness of the equipment.

Third, front - line data loading and combined application planning. According to operational tasks and combined with operational data, combat troops make operational plans and give full play to the maximum effectiveness of their equipment through effective technical and tactical coordination.

Fourth, strategy adjustment and loading. Field reprogramming of movable combat support equipment is used to adjust the equipment loading data in the light of battlefield emergencies.

Fifthly, re-enactment analysis of all combat operations after the war. Analyze the effectiveness of confrontation and the rationality of equipment application, extract operational experience data, optimize and improve the data model and countermeasures.

3 Revelations and suggestions

3.1 Constantly improve the electronic countermeasures equipment system centering on mission expansion and equipment technology development

In view of the new demand for electronic countermeasures equipment, such as the transformation and development of the Navy, especially the far-sea mobile operations of the Navy's aircraft carrier formation, the defense of nansha islands and reefs, the deployment of overseas bases and the safeguarding of maritime rights and non-war military operations, etc. In the aspect of electronic reconnaissance, the emphasis should be given to the comprehensive processing of regional multi-source information and the networked application of individual identification of radiation sources. In the aspect of electronic attack, efforts should be made to construct the jamming system of maritime space-air information system and the jamming system of military firepower penetration support to comprehensively improve the electronic attack capability. In the field of electronic defense, the countermeasures should be supplemented to improve the ability to counter new threats, and electronic defense equipment should be developed systematically to meet the needs of ports, airports and other electronic defense.

https://doi.org/10.1051/shsconf/20219601012

3.2 Strengthen the study of main combat adversaries and improve the effectiveness of electronic countermeasures

Different from radar and other sensor equipment, missile and other weapon systems, electronic countermeasures require specific targets, accurate timing, reasonable means and matching jamming styles, so as to achieve targeted reconnaissance jamming and achieve better countermeasure effectiveness. Therefore, it is not only necessary to fully grasp the technical characteristics of combat objects, such as parameter characteristics and working mode, but also to study their tactical characteristics, such as equipment performance, activity rules and application process, so as to effectively change the current situation that electronic countermeasures are not highly targeted and combat effectiveness is low.

References

1. J Guo, 60 cases of Electronic Warfare [M]. Beijing: PLA Press (2006)
2. L Lei, X Shi, LV Zeejun, et al. Reconnaissance and Surveillance - Clairance-Eye and Wind-Drift Ear in combat Space [M]. Beijing: National Defense Industry Press (2012)
3. Y Wang, S Zhu. International Electronic Warfare [J]. Electronic Information Control Laboratory (2017)
4. Y Mao. Analysis on improvement project of electronic warfare system of us navy surface ships [J]. Ship electronic engineering, **30**, (2010)
5. Q L Xiong, Integrated Electronic Warfare -- The Killer Mace of information War [M]. Beijing: National Defense Industry Press (2008)
6. K J Liu. American Future Combat System [M]. Beijing: PLA Press (2006)

Research on the Influence Mechanism of Career Insecurity on Individual Career Dynamics

Peiqi Wang *, *Bin* Zhang and *He* Huang

University of Defense Technology, No.5 Guang Ming Road, Xi'an, Shanxi, China

Abstract. Career insecurity is widespread in individual career development. Career insecurity has both positive and negative effects on individual career dynamics. It has practical value to explore the influence mechanism between career insecurity and individual career dynamics for both individual career development and organizational performance. Through sorting out the relevant literature, this paper proposes two hypotheses on the influence of career insecurity on individual career dynamics. The negative impact of career insecurity on individual career motivation is mainly reflected in the reduction of organizational commitment, organizational trust, job satisfaction, employee performance, work input, innovative behavior, etc.; the positive impact of job insecurity on individual career motivation is mainly reflected in a low level of career insecurity can lead to social inertia such as absenteeism and increased escapism. this research plans the research methods and steps, laying a foundation for the next empirical study.

Keywords: Career Insecurity, Career Dynamics, Influence Mechanism

1 Introduction

The theory of career dynamics is also called career development view, which was first put forward by professor E.H. Schein[1]. Its essence is the interaction between the inner occupation (individual) and the outer occupation (organization), and the mutual cooperation, so that the needs of both sides can be satisfied, benefit from each other and develop. The narrow sense of career dynamics refers to the motivation of individual work. It is generated by a variety of related factors inside and outside individual career, which urges individuals to engage in work activities, and promotes the operation and development of work activities cycle by cycle, so as to achieve the work objectives of the countless sub forces formed by the resultant force. Internal motivation is composed of needs, interests, wills, values and world views, while external motivation is formed by factors such as social status, economic benefits and working conditions. Only when external forces are transformed into internal forces can they play a positive role. This study mainly focuses on the narrow sense of career dynamics, which refers to the integration of factors that can stimulate individuals to engage in work, develop themselves and realize organizational needs at the same time. Most of the researches on career dynamics are conducted from the perspective of a specific group's professional identity or career plateau. For example, a study on the professional identity of teachers (li

* Corresponding author: 463872349@qq.com

Peng, a study on the motivation of teachers' career, 2010; Xue Yijun, Research on career dynamics of university counselors based on expectation theory, 2015). Careers insecurity is a subjective psychological feeling, which has an important influence on the individual's attitude and behavior. In general, career insecurity is a major source of career stress for individuals. The individual feels the career insecurity and falls into the negative emotion, which causes the negative behavior in the work and affects the career dynamics. There are many researches on individual negative work behaviors in the research field. For example, the degree of career satisfaction decreases, career involvement decreases, and career burnout increases. However, there is little research on the influence of career insecurity on individual career dynamics. The research on the influence mechanism of career insecurity on individual career dynamics can provide reference for the human resource management of organizations.

2 Basic Theories

2.1 Career insecurity

Scholars began studying career insecurity in the 1980s. According to scholars, career insecurity is the feeling of powerlessness caused by employees who are in a threatened work environment but still want to maintain their current work. They also point out that this kind of uncertainty comes from multiple directions. As long as employees face any job-related loss, such as loss of income, promotion opportunity, career development opportunity, status and colleagues, they will feel insecure about their jobs. Some scholars define it as "the subjective perception of the individual when the continuity of the present job is threatened", "It's the individual's fear of the threat of not being able to continue their current job." In general, these definitions of career insecurity focus on whether you can continue in your current job in the future. However, there are still some disputes over the definition of the category and no unified cognition has been formed. At present, the connotation of career insecurity can be summarized as follows:

Firstly, career insecurity is subjective and objective. Researchers in the field of organizational behavior tend to regard career insecurity as an objective situational phenomenon, independent of individual experience and interpretation. In other words, no matter how the individual perceives it, Career insecurity can be evaluated objectively and independently of the individual. For example, there are objective differences in career insecurity between regular employees and temporary employees. The form of temporary employment itself is an objective existence of career insecurity. Some researchers also believe that career insecurity is a subjective psychological phenomenon, which is the expectation of subordinates that their job survival will be threatened. It is full of subjectivity.

Secondly, career insecurity can be acute or chronic. Acute insecurities such as urgent job stressors, such as sudden mass layoffs, can trigger a sense of crisis. Chronic insecurity is a more common feeling, such as anxiety caused by high unemployment in the workforce, so career insecurity exists not only in organizations that are in a period of change, but also in organizations that don't seem threatening.

Thirdly, the career insecurity has the whole and the multi-dimensional distinction. Some scholars believe that career insecurity is a one-dimensional structure, which is the overall consideration of employees on whether to continue their current work. Some scholars also believe that career insecurity is multi-dimensional, including not only the fear of losing the job opportunity, but also the fear of the value characteristics brought by losing the job.

2.2 Factors influencing occupational motivation

Career dynamics theory rejects the idea that the organization decides the demands of the career unilaterally. Instead, it believes that career development is realized by the process that the inner career and the outer career work together, and the inner career and the outer career match each other. An individual's career dynamics should consist of two major elements: first, the individual's internal drive, also known as the career anchor, which is subjective and includes values, interests, skills, etc.; The second is external driving force, such as industry classification, occupational orientation, social factors, etc., specific to the organizational level, including recruitment, training, promotion, compensation, punishment and other mechanisms. It is important to emphasize that the two do not exist in isolation, but interact with each other and eventually the external drive ACTS through the internal drive.

The career dynamics we discuss here is based on the premise that the career choice has been made clear, that is, a certain occupation has been selected, and then to explore what factors within this occupation can stimulate the individual to continue to maintain the confidence and driving force of the vigorous career development. In fact, he prefers the concept of psychological contract, that is, the expectation that exists between the organization and the individual that cannot be written down. Based on the concept of psychological contract, the factors affecting career dynamics can be summarized as follows: good working environment, safety and sense of belonging, promotion, training and development opportunities, value identification, remuneration, and the coincidence between task and career orientation, etc. Some scholars also equate occupational motivation with occupational motivation and define it as three dimensions: self-efficacy beliefs, person-job fit, and work engagement.

3 Relational Hypothesis

3.1 Existing research conclusion

The research on the influence mechanism of job insecurity on individual occupational motivation can be expressed by a model. Starting from the individual's psychological perception to work insecurity, exploring the impact on occupational motivation and its mechanism of action, as far as the work itself is concerned, how the complexity of job requirements affects the relationship between work safety and individual occupational motivation. As shown in Figure 1:

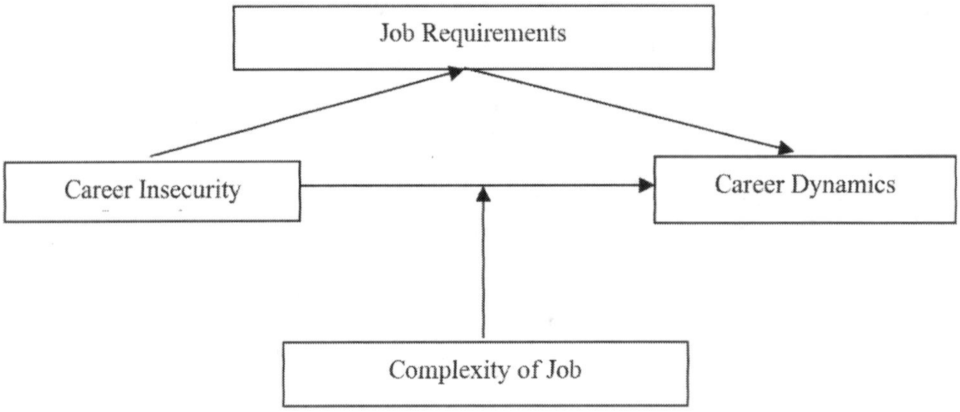

Fig. 1. The Model of The Effect of Job Insecurity and Individual Occupational Motivation.

From the perspective of the mechanism of organization identification factors and work complexity factors in Figure 1 (according to existing research conclusions [2]), in fact, the impact of job insecurity on individual occupational motivation cannot be simply used to linear relationship It is said that there are many intermediate variables, such as organizational identity and work complexity as shown in the figure. Therefore, to study the mechanism of job insecurity on individual occupational motivation, it is necessary to first select the intermediary factors that have an impact on it.

You can refer to the existing research conclusions, such as: the job content and job requirements of different positions will directly affect the employee's work input level; under the Chinese background scenario, there is a significant negative relationship between job insecurity and organizational identity; There is a significant positive correlation between job safety and job engagement; organizational identity plays a partial intermediary role in the relationship between job safety and job engagement. From the data results, organizational identity cannot fully explain the relationship between the two. There are other possible explanations; job complexity can effectively regulate the relationship between the two, that is, when job complexity is at a low level, there is a significant positive correlation between job safety and job engagement; when job complexity level is relatively low At high times, there is a significant positive correlation between job safety and job engagement, and when the relative job complexity level is low, the positive relationship between job safety and job engagement is more pronounced [6].

According to existing research conclusions, job trust also has an impact between job insecurity and individual career motivation [7]. Work trust reflects the individual's willingness or belief to believe in and depend on others, and at the same time reflects the individual's willingness to take risks for the trusted party [8]. Employee's trust in the organization is a kind of employee's overall perception of trust in the organization as a whole. It is usually based on the employee's decision and action on the organization's leadership, as well as the organization's governance mechanism and incentive system to form the overall impression of the organization [9]. At work, employees rely on their organizations to provide them with a good working environment, reasonable salaries, and future growth opportunities, which forms employees' trust in the organization [6]. The increase in job insecurity will reduce employees' trust in the organization, which in turn will reduce employees' creative process investment and intrinsic motivation, impair their creativity performance, and then have a negative impact on individual career motivation.

3.2 Hypothesis of negative impact of career insecurity on occupational motivation

Since career insecurity is an organizational member's psychological expectation of future job sustainability, career insecurity will obviously have an impact on the individual's psychology and then have an impact on his/her behavior. A large number of empirical studies show that: high level of career insecurity can reduce organizational commitment, organizational trust, job satisfaction, employee performance, job involvement, innovation behavior, etc. This is a negative effect of career insecurity. The research in this aspect mostly USES the method of empirical research. For example, the study of Hultgren (2003) shows that career insecurity has a negative impact on motivation and organizational commitment. Holmgren, Sverker and Isaksson (1999) conducted a longitudinal study of 375 employees in a shrinking organization in Sweden and found that career insecurity has a negative impact on organizations and individuals. However, the empirical study on the relationship between career insecurity and job performance found that the relationship between the two was very complex, and Rosenblatt et al. (1999) found that the two were negatively correlated. At the same time, Wong et al. (2005) investigated the impact of career insecurity on organizational citizenship behavior and job performance in the context of China, and found that the impact depended on the nature of the enterprise and the trust of employees, etc. Therefore, it can be confirmed that career insecurity has a negative impact on individual career dynamics.

3.3 Hypothesis of positive impact of career insecurity on career dynamics

On one hand, career insecurity has a negative impact on individual career dynamics; on the other hand, empirical studies have shown that if work security is at a high level, it will have a negative impact on performance. For example, studies have shown that "low levels of career insecurity lead to increased absenteeism and escapism."[2] In fact, the findings can be found in classic theories of organizational behavior, such as "social loafing" in group behavior. Due to the diffusion of responsibilities, for example, in the performance evaluation, there is a lack of targeted assessment for individuals and a lack of differentiation in the distribution of benefits, which leads to a higher level of job security for individuals. Therefore, performance is reduced. Career insecurity should be in a balance, moderate career insecurity can stimulate the career dynamics of individuals, so that they can work hard and improve themselves.

4 Empirical Design

Research can use questionnaire survey method, select more than 20 teams to collect data. To avoid deviations in common methods, questionnaires can be filled out by employees and their team leaders. When investigating, first determine the team that needs to participate in the survey with the assistance of the human resources department of the enterprise, and obtain the list of team leaders and members. Then number the employees, and mark the number on the employee questionnaire and issue it to the employee corresponding to the number to fill out. At the same time, the name of the employee to be evaluated is marked on the leadership questionnaire and distributed to the corresponding team leader to fill in. Each leader may need to evaluate the professional motivation of multiple team members. After the questionnaire was recovered, the employee questionnaire and the leader questionnaire were matched according to the list number. Questionnaire respondents. The research can adopt the questionnaire survey method to select the individuals in specific organizations who are faced with the situation of post change adjustment, organizational re-organization or regional change for investigation. The selection of questionnaire can draw on the existing mature scale, such as the scale carried out by Lee eta in the context of China. [3] After the scale is selected, the initial modification can be made according to the survey object, and the modification method can be carried out with the method of expert review. After the first round of distribution, further screening of the question options can be conducted by using SPSS statistical software and factor analysis method to determine the scale questions. In the process of empirical design, the selection of scale is very important, and the particularity of occupation and the rationality of samples should be fully considered.

5 Conclusion

Through the research of the above methods, the following conclusions can be drawn: First, job insecurity has an impact on individual occupational motivation, and this effect is not simply a positive or negative correlation. Instead, there must be some intermediate variables. According to the existing research conclusions, these intermediate variables may be organizational identity, work complexity, organizational trust, employee creativity, etc.; second, they can make job insecurity feel negative for individual career motivation. Hypothesis of impact: Work insecurity has a negative impact on personal professional motivation by reducing organizational commitment, organizational trust, job satisfaction, employee performance, work input, innovative behavior, etc. Third, a certain amount of work insecurity can be made Individual career motivation has a certain role in promoting, that is, the assumption that it has a positive impact on individual career motivation can enable individuals to maintain a more positive attitude into work. After verifying the hypothesis, we can find the correlation between job insecurity and individual occupational motivation, and then ensure the subordinates' work safety in organizational management while avoiding excessive work insecurity and work safety. Reach a relative balance. On the one hand, it can stimulate the internal motivation of employees and make them actively face the work; on the other hand, it can also ensure a certain sense of crisis and jump out of the "comfort zone" to meet more challenges and changes and stimulate their creativity. Conclusion After verification, to verify the mechanism of job insecurity on individual occupational motivation, it is also necessary to select typical samples and conduct empirical research in conjunction with statistical analysis.

Acknowledgements

This work is supported by "2018 National Social Science Foundation of China 18BGL291".

References

1. Y Sun. The internal mechanism of occupational motivation theory: based on the perspective of value chain. Enterprise Technology Development, **3**, 99 (2008)
2. C X Liang. Research on the impact of employees' job insecurity and psychological empowerment on work input, **8** (2016)
3. A Biletm. A Model of Creativity and Innovation in Organizations. Research in Organizational Bahavior, **10**(1) 123-167, (1988)
4. Y L Zhao, The mediating role of job insecurity in the relationship between employee self-perceived employability and psychological well-being. Tianjin Normal University. **5**, (2018)
5. N Yao. Research on the relationship between job insecurity and job satisfaction based on organizational support. Zhejiang University of Technology, 6 (2012)
6. L Huang. Research on the influence mechanism of work safety on employees' work input. Knowledge Economy, **19** 111 (2019)
7. Z W Li, L H Chen, F R Wang. Research on the impact mechanism of job insecurity on employee creativity. Science and Technology and Economy, **10**, 76-78 (2019)
8. J Leekc, L Hahnm. Empirical analysis of roles perceived leadership styles and trust on team members' creativity: Evidence from KoreanI CT companies Computers in Human Behavior, **42**(C),149-156 (2015)
9. R C Nyhan, H A Marlowe Jr. Development and psychometric properties of the organizational trust inventory. Evaluation Review, **21**(5):614-635, (1997)
10. X T Zheng, J T Shi Jintao, Zheng Xingshan. The influence of employee trust in their organization on their work attitude. Management Review, **20**(11), 36-40 (2008)

Job Burnout and Its Relevant Factors among Medical Interns

Yongmei Hou* *Haojun* Fu

Department of Psychology, School of Humanity and Administration, Guangdong Medical University, Dongguan 523808, Guangdong Province, China

Abstract. Objective: to understand the present situation of job burnout among medical interns and to explore its major influencing factors. Methods: A stratified random sampling was used to select 1268 medical interns from 6 full-time medical colleges in Guangdong province. They were investigated with Maslach Burnout Inventor-General Survey, (MBI－GS), Questionnaire on Humanistic Care Ability of Medical Students (QHCAMS) and a self-compiled personal information questionnaire. Results: (1) The incidence of job burnout in this group was 98.40%. The total score of MBI－GS and the scores of emotional exhaustion, cynicism and reduced personal accomplishment were (45.27 ±12.61), (16.21 ±4.91), (12.91 ±3.93) and (18.15 ±5.22), respectively. (2) The result of multiple stepwise linear regression showed that gender, specialty, type of internship hospital, working department and average length of overtime per day were positively correlated with the total score of MBI-GS (β =.154～.676, all P < 0.01). Nine factors, such as self-evaluation of health status, ability to solve problems at work, relationship with patients, classmates and medical staff, as well as the scores of hope, honesty, humility and courage in HQCAMS, were negatively correlated with the total score of MBI-GS (β =-.148～-.468, all P < 0.01). Conclusion: Medical interns are suffering serious job burnouts, which may be related to work pressure, personal quality and other factors.

Keyword: Medical Interns, Job Burnout, Relevant Factors

1 Introduction

The internship is an important transitional stage for medical students to change from a student to a clinical practitioner. Successful adaptation during this stage is of great significance for medical students to establish a proper professional attitude, strengthen professional aspiration, improve work enthusiasm and work efficiency, and promote career development. However, medical interns are found to suffer great mental stresses and are prone to job burnouts because of their incomplete and immature professional skills, insufficient psychological preparation and incompetence in work. Researches showed that the incidence of job burnout among medical interns in China is more than 70.9% [1-4]. With regard to the influencing factors of job burnout, previous studies have mostly focused

* Corresponding author: 2184456621@qq.com

on a limited number of demographic factors, but comprehensive studies on the combination of demographic factors and psychosocial factors are few.

Based on the above analysis, this study intends to adopt a large sample and multi-center investigation model to make a comprehensive analysis of this problem from many aspects, such as demographic factors, psychosocial factors, and so on.

2 Objects and methods

2.1 Objects

Stratified random sampling was used to select medical interns from 6 full-time colleges (Southern Medical University, Guangzhou University of Traditional Chinese Medicine, Guangdong Medical University, Guangzhou Medical University, Guangdong Pharmaceutical University and Shenzhen University Health Science Center) as the study objects. 1300 questionnaires were sent out and 1268 valid questionnaires were collected, with an effective rate of 97.54%. Among them, 1237 were from Han nationality, 31 from ethnic minorities, 233 from Southern Medical University, 345 from Guangzhou University of Traditional Chinese Medicine, 274 from Guangdong Medical University, 138 from Guangzhou Medical University, 156 from Guangdong Pharmaceutical University, 122 from Shenzhen University Health Science Center; 1094 are only-child, 174 are non-only-child; 2 from rich family, 652 from well-off family, 523 from poverty-line family, and 72 from needy family; 763 from urban areas, 505 from rural areas; 46 academically excellent, 636 academically moderate, 495 academically intermediate, 88 passing score and 3 failed.

2.2 Tools

2.2.1 Maslach Burnout Inventory —General Survey (MBI —GS)

It was compiled by Maslach et al. (1984) and revised by Li Chaoping to the Chinese version [5]. There were 15 items, divided into 3 dimensions: emotional exhaustion, cynicism and reduced personal accomplishment. The Likert 7-point scoring method is used to score from 0 (never)-6 (daily). The higher the score, the more obvious the burnout tendency. The critical values of emotional exhaustion, cynicism and reduced personal accomplishment were 25 points, 11 points and 16 points, respectively. If all the three dimensions < the critical value, the objects would be considered not suffer burnout; if one of the three dimensions ≥ critical value, the object would be considered suffer mild burnout; if two of the three dimensions ≥ critical value, the object would be considered suffer moderate burnout; and if all of the three dimensions ≥ critical value, the objects would be considered suffer serious burnout. The average score of the scale (dimension) below 3 indicates low job burnout, 3-5 indicates serious job burnout, and more than 5 indicates very serious job burnout. In this study, the Cronbach α coefficient of the total scale was 0.89, and the Cronbach α coefficient of the three dimensions was 0.80-0.84.

2.2.2 Questionnaire on Humanistic Care Ability of Medical Students (QHCAMS)

It was compiled by Shen Zhengfu el. (2013) [6], involving 42 items and 7 dimensions, namely, altruism, honesty, courage, trust, patience, humility and hope. The Likert 5-point scoring method is used to score from 1(not at all) to 5 (completely yes). The higher the total score, the more obvious the trendency in this dimension (item). In this study, the

Cronbach'a coefficient of the total questionnaire was 0.85, and the Cronbach α coefficient of 7 dimensions was 0.77-0.83.

2.2.3 Self-compiled Personal Information Questionnaire

The CNKI, Wanfang database, VIP database, Baidu, Pubmed and other search engines were used to search the literatures about the burnout among medical interns (277 in Chinese and 2421 in foreign languages). Based on that, the basic contents of the questionnaire were constructed, with a total of 19 items. Combined with the results of 3 collective discussions with 10 representatives of medical interns and 5 experts in the field of higher medical education, 4 items were deleted and 3 items were added. It involves 18 items, which include gender, age, nationality, school, specialty, family status (only-child or not), family financial status, origin, academic achievement, type of internship hospital, working department, length of overtime every day, ability to solve problems at work, relationship with patients and medical staff and classmates, self-assessment of personality and health status and so on.

2.3 Data manipulation

SPSS 20.0 was used for statistical analysis. Descriptive statistics were used to calculate the average score and standard deviation of each scale; independent sample t-test, Pearson product correlation and one-way ANOVA were used to explore the correlation between variables; multiple stepwise linear regression was used to analyze the related factors of MBI-GS total score.

3 Results

3.1 Descriptive statistics of the total score and factor score of each scale

The incidence of job burnout in this group was 98.40%. Among them, there were 946 (74.6%) with mild burnout, 238 (18.5%) with moderate burnout and 67 (5.3%) with serious burnout.

As can be seen from Table 1, the overall level of job burnout in this group is on the high side (MBI-GS total score = 3.95 ±0.84).

Table1. Descriptive statistics of the scores of each scale (n=1268).

Dimensions	Min	Max	M	SD	Number of items	Average score of each item	Standard deviation of each item
total score of MBI-GS	35.00	87.00	45.27	12.61	15	3.95	0.84
emotional exhaustion	5.00	31.00	16.21	4.91	5	4.04	0.98
reduced personal accomplishment	6.00	31.00	18.15	5.22	6	3.86	0.87
cynicism	5.00	24.00	12.91	3.93	4	3.98	0.99
total score of QHCAMS	91.00	210.00	144.74	18.98	42	3.44	0.45
altruism	7.00	25.00	16.86	3.09	5	3.37	6.17
honesty	7.00	25.00	17.26	3.61	5	3.45	0.72
courage	7.00	25.00	16.34	2.97	5	3.27	0.59
trust	13.00	40.00	27.61	4.56	8	3.45	0.57
patience	13.00	35.00	24.72	4.58	7	3.53	0.65
humility	11.00	35.00	24.46	4.60	7	3.50	0.66
hope	5.00	25.00	17.49	3.86	5	3.44	0.45

3.2 Correlation Analysis between the scores of each scale

From Table 2, it can be found that there is a significant pairwise correlation between the total score of MBI-GS (together with the scores of 3 dimensions) and the total score of QHCAMS (together with the scores of 7 dimensions).

Table 2. Analysis of Correlation between MBI-GS and QHCAMS Scores (n=1268).

Dimension	1	2	3	4	5	6	7	8	9	10	11	15
1MBI-GS	1											
2 emotional exhaustion	.884**	1										
3 reduced personal accomplishment	.930**	.719**	1									
4 cynicism	.870**	.630**	.756**	1								
5 altruism	-.170**	-.156**	-.140**	-.165**	1							
6 honesty	-.313**	-.214**	-.297**	-.341**	.302**	1						
7 courage	-.341**	-.311**	-.298**	-.052	.164**	.183**	1					
8 trust	-.242**	-.196**	-.199**	-.268**	.299**	.398**	.239**	1				
9 patience	-.294**	-.217**	-.260**	-.326**	.388**	.482**	.253**	.496**	1			
10 humility	-.334**	-.231**	-.310**	-.370**	.300**	.471**	.252**	.526**	.562**	1		
11 hope	-.377**	-.312**	-.317**	-.398**	.304**	.461**	.150**	.492**	.503**	.558**	1	
12 HQCAMS	-.370**	-.276**	-.325**	-.410**	.553**	.692**	.437**	.746**	.793**	.794**	.724**	1

Note: * P<0.05， **P<0.0

3.3 Multiple stepwise linear regression analysis of factors related to job burnout among medical interns

3.3.1 Variable Assignment

First, values are assigned to the possible situations (alternative answers) of demographic and psychosocial classification variables that may affect the total score of MBI-GS, and the results are shown in Table 3.

Table 3. Variable Assignment.

item	Options and variable assignment
1.nation	0=minority minorities; 1= Han nationality
2.gender	0= male;1= female
3.school	0=Southern Medical University; 1=Guangzhou University of Traditional Chinese Medicine; 2=Guangdong Medical University; 3=Guangzhou Medical University; 4=Guangdong Pharmaceutical University; 5= Shenzhen University Health Science Center
4.specialty	0=Oral Medicine; 1= Preventive Medicine; 2= Medical Examination; 3= Pharmacy; 4= Nursing; 5=Anesthesiology; 6= Clinical Medicine
5. From only-child family or not	0=no; 1=yes
6. Family financial status	0= rich; 1= moderate; 2= poverty-line; 3= poor
7. origin	0= urban area; 1= rural areas
8. academic performance	0= failed; 1= passed; 2= medium; 3= moderate; 4= excellent
9. Type of internship hospital	0=fiest class; 1=second class; 2=tertiary
10.working department	0= Department of Internal Medicine; 1= Department of Obstetrics & Gynecology; 2= Department of Pediatrics; 3= Surgical Department; 4= Department of psychiatry; 5= Department of Infectious Diseases;6=ICU
11. Average length of overtime per day	0=0 hour; 1=within 1hour; 2=1-2hours; 3=2-3hours; 4=More than 3 hours
12. Ability to solve problems at work	0= Almost can't; 1= Mostly can't; 2= Mostly can; 3= Totally can
13. Relationship with patients	0= Not sure; 1=Bad; 2= Moderate; 3=Good; 4= Great
14. Relationship with medical staff	0= Not sure; 1=Bad; 2= Moderate; 3=Good; 4= Great
15. Relationship with classmates	0= Not sure; 1=Bad; 2= Moderate; 3=Good; 4= Great
16. self-assessment of personality	0= extraversion， 1= introversion

17. Self-assessment of health status	0= not well; 1= moderate;2= strong

3.3.2 Multiple stepwise linear regression analysis of factors related to job burnout among medical interns

Taking the total score of MBI-GS as the dependent variable, the 7 dimensions of QHCAMS and 18 factors that may be related to the total score of MBI-GS (including demographic variables and psychosocial factors) as independent variables, a multiple stepwise linear regression was carried out within 95% confidence interval.

From Table 3, it can be found that gender, specialty, type of internship hospital, working department and average length of overtime per day were positively correlated with the total score of MBI-GS (β =.154～.676, all P < 0.01). Nine factors, such as self-evaluation of health status, ability to solve problems at work, relationship with patients, classmates and medical staff, as well as the scores of hope, honesty, humility and courage, were negatively correlated with the total score of MBI-GS (β =-.148～-.468, all P < 0.01).

Table 4. Multiple stepwise linear regression analysis of factors related to job burnout among medical interns.

Dependent Variable	Independent variables	B	SE	β	t value	P value	R^2	R_{adj}^2
MBI-GS Total score	Gender	.267	.045	.154	7.623	<.001	.508	.503
	specialty	.381	.069	.312	2.999	.005		
	Type of internship hospital	.427	.116	.374	2.766	.008		
	working department	.683	.144	.599	3.863	<.001		
	average length of overtime per day	.485	.132	.402	4.899	<.001		
	self-evaluation of health status	-.313	.051	-.302	-5.209	<.001		
	ability to solve problems at work	-.753	.105	-.676	-2.785	.005		
	relationship with patients	-.440	.196	-.386	-4.464	<.001		
	relationship with medical staff	-.528	.213	-.468	-5.095	<.001		
	relationship with classmates	-.344	.092	-.284	-3.900	<.001		
	hope	-.257	.055	-.236	-4.706	<.001		
	honesty	-.173	.055	-.148	-3.132	.002		
	humility	-.202	.066	-.158	-3.084	.002		
	courage	-.224	.043	-.191	-5.116	<.001		

4 Discussion

This study found that medical interns suffered a rather high level of job burnout, with the average MBI-GS score being 3.95±0.84, 98.40% of the medical interns suffered from job burnout, which was consistent with the results of previous studies [1-4]. The result suggests that job burnout is a common psychological problem among the medical interns in China.

The results of multiple stepwise linear regression showed that gender, specialty, type of internship hospital, working department and average length of overtime per day were positively correlated with the total score of MBI-GS. Nine factors, such as self-evaluation of health status, ability to solve problems at work, relationship with patients, classmates and medical staffss, as well as hope, honesty, humility and courage, were negatively correlated with the total score of MBI-GS.

Girls suffer higher level of job burnout than boys, which is consistent with the findings of Shu Xianzhu [4]. It is suggested that there are gender differences in occupational adaptation. Job burnout is a negative mood and behavioral reaction caused by the inability to tackle the stress at work. Besides, girls are less capable in anti-frustration and emotion regulation than boys [7]. Stress will bring about greater and more lasting stress on girls, resulting in excessive activation of negative emotions [8]. Therefore, it is concluded that girls are more likely to suffer job burnout.

It is also found that specialty, the type of hospital, the working department, the average length of overtime per day are positively correlated with the total score of MBI-GS, which is consistent with the research results of Shu Xianzhu [4]. The above 5 factors reflect the objective work stress from different angles, such as work task, work content, work standard, workload and so on, which have positive predictive effects on the job burnout of the enrolled medical staff. The result has been confirmed by a large number of literatures [9-14]. Once again, this study confirmed that the results also have a positive predictive effect on the job burnout of medical interns, suggesting that objective job stresses have a generally positive predictive effect on job burnout among medical workers.

Lastly, it was found that such nine factors as self-evaluation of health status, ability to solve problems at work, relationship with patients, classmates and medical staff, and QHCAMS's hope, honesty, humility and courage, were negatively correlated with the total score of MBI-GS. The result is consistent with the findings of previous studies [4,15], suggesting that good personal qualities (physical health status, ability to solve problems at work, personality traits, interpersonal relationships, etc.) are helpful to reduce the job burnout among medical interns.

References

1. H Li, B R Liu, X M Liu, et al. *Initial investigation about occupational lassitude condition of clinical interns and nursing students,* China Higher Medical Education, **7**, 62-64 (2011)
2. H Song, W Xu. *A Study on the relationship between achievement motivation and job burnout of medical interns*, China Higher Medical Education, **5,** 62-64 (2015)
3. L J Yang, Q L Peng, X N Zhou. *Analysis of job burnout situation in TCM medical interns from perspective of emotion regulation self-efficacy*, Journal of Changchun University of Chinese Medicine, **31**(6), 1288-1290 (2015)
4. X Z Shu, T D Huang, N C Fan, et al. *Correlation of occupational burnout with social support and copying style among interns of a medical college in Sichuan*, Pract Prev Med, **23**(11), 1340-1344 (2016)

5. C P Li, K Shi. *The influence of distributive justice and procedural justice on job burnout*, Acta Psychologica Sinica, **35**(5), 677-684 (2003)
6. Z F Shen, X M Yang, H X Han, et al. *Preparation and application of questionnaire on humanistic care ability of medical students*, Journal of the Third University, **15,** 77-80 (2013)
7. M R Wang, L Zeng. *An empirical study on the psychological characteristics of post-90s female college students -- Based on the comparison with male college students*, China Youth Study, **4,** 67-72 (2014)
8. Y H Hu, M Huang, *Diverse consequences of neuroticism and extraversion on down_regulation of negative emotions,* ctaPsychologicaSinica, **38**(4), 553-561 (2006)
9. R Gashmard, R Bagherzadeh, S Pouladi, et al. *Burnout and its related demographic factors among the medical staff working in hospitals associated with Bushehr University of medical sciences*, Puerto Rico Health Sciences Journal, **34**(4) (2015)
10. X Y Zhou, J C Pu, X Zhong, et al. *Burnout, psychological morbidity, job stress, and job satisfaction in Chinese neurologists*, Neurology, **88**(18), 1727- 1735 (2017)
11. L Y Zhang, J M Zhang, Y F Gao, et al. Comparative Study on the Status of Doctors ' and Nurses ' Job Burnout and Its Influence Factors, China Journal of Health Psychology, **19**(11), 1345-1347 (2011)
12. Hao Cheng, H Ma, N Zhang. *The status of occupational burnout among doctors in west China and its relationship with work-family conflict*, Chinese Journal of Behavioral Medical Science, **21**(3), 262-264 (2012)
13. J M Zhang, J Q Sun, Q L Liu, et al. *Investigation on Job Burnout of nurses in general hospitals*, Journal of Ningxia Medical University, **37**(2), 216-218 (2015)
14. C B Wu. *Research on Occupational Burnout and Resilience of Psychiatric and Non-psychiatric Medical Staff, Ma.D. Thesis*, Jining Medical University (2019)
15. X Feng, H T Li. Research on quality education of medical students based on the prevention of job burnout, NORTHWEST MED.EDU, **22**(5), 929-930 (2014)

A Study on Satisfaction of Elderly Care Service Quality Based on Different Groups: Taking the Monitoring Data of Liaoning Province as an Example

Jingxing Liao*, *Lei* Feng, and *Juxiu* Huang

China National Institute of Standardization, China

Abstract. The monitoring and analysis of the social public service quality is conducive to promoting the construction of a service-oriented government that the people are satisfied with. In this paper, the third-party quality monitoring is carried out in 14 cities in Liaoning Province from 11 public service areas, including residential environment, public transportation, infrastructure, medical and health care, public security, cultural and sports, compulsory education, elderly care service, employment service, social security and administrative convenience. The results show that the public's satisfaction with social security, compulsory education, residential environment, public security and employment service is low, which needs to be paid attention to and further improved by Liaoning provincial government.
Keywords: Public Service, Quality Monitoring, Investigation and Analysis.

1 Introduction

With the rapid development of China's economy and society and the decisive stage of building a well-off society in an all-round way, more and more elderly people and their children are in need of high-quality elderly care service. The construction of elderly care service quality will become an important part of our country's economic development, social progress and improvement of people's livelihood for a long time in the future[1]. Providing effective high-quality elderly care service is an important direction for the development of elderly care service, and how to improve and perfect the elderly care service policy, etc. has become a key issue for the government to deal with the elderly care service quality problem[3]. In order to find out the advantages and disadvantages of the elderly care service quality, to put forward some precise suggestions, to break down development barriers through opening up "blocking points" and eliminating "pain points", and to promote the improvement of the elderly care service quality, this paper monitors the quality of elderly care services in Liaoning Province from six areas, including public elderly care services, community elderly care services, home-based elderly care services, resettlement for the

* Corresponding author: liaojx@cnis.ac.cn

elderly, preferential policies and services for the elderly population, and public elderly care services[2] [4] [5].

2 General situation of elderly care service quality in liaoning province

The monitoring results show that in 2019, the overall evaluation level of the public satisfaction survey results of the elderly care service in Liaoning Province is "relatively satisfactory", with a score of 67.93, an increase of 0.66 compared with last year. The public's evaluation on the measurement indexes of the elderly care service is higher than that of 2018. In 2019, there are significant differences in the public's satisfaction scores in the various fields involved in the elderly care service. The public's overall feelings about the elderly care service quality and the home-based elderly care service quality score are higher, with the score of 69.33 and 69.20, respectively. The second is the resettlement of the elderly, the preferential policies and services for the elderly, the community's elderly care service, the public elderly care service and the local elderly care service cost. The scores are 68.35, 67.86, 67.60, 67.21 and 67.19 in turn. The service quality score of local elderly care institutions is the lowest with 66.72. The scores of each index were shown in Figure 1.

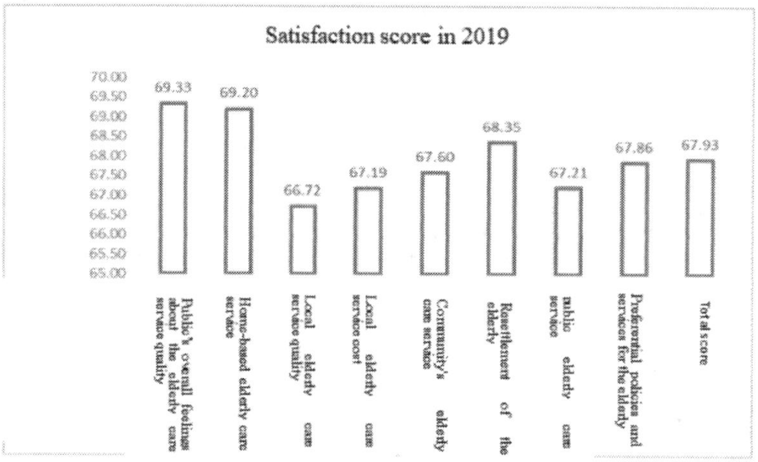

Fig.1. The public's satisfaction with the elderly care service.

3 Analysis on the elderly care service quality in different groups

3.1 Quality of elderly care service in different age groups

Statistics show that among the respondents who are willing to reveal their age in 2019 elderly care service quality satisfaction survey, there are 2,250 young people (aged 18-44), accounting for 61.78% of the total; 1,130 middle-aged people (aged 45-59), accounting for 31.03% of the total; and 262 elderly people (aged 60-79), accounting for 7.19% of the total. The data are shown in Figure 2 and Table 1.

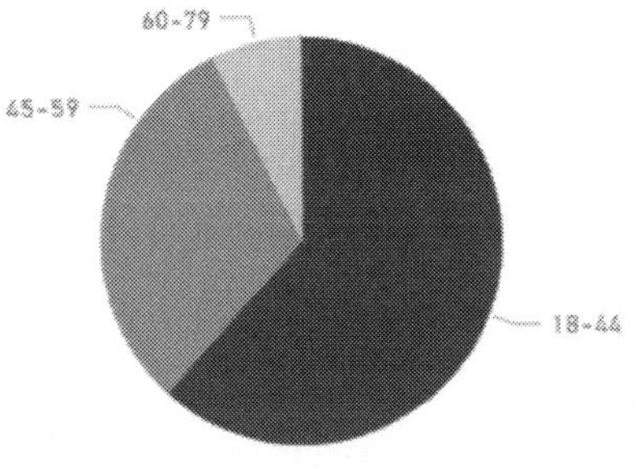

Fig. 2. The Age Proportion of the Respondents in the Social Satisfaction of the Elderly Care Service Quality in 2019

Table 1. The Age Proportion of the Respondents in the Social Satisfaction of the Elderly Care Service Quality in 2019.

Age group	Age group count	% GT of age group count
18-44	2250	61.78%
45-59	1130	31.03%
60-79	262	7.19%

There is a small difference in the satisfaction of the elderly care service among different age groups. Among them, the satisfaction of the elderly group and the young group to the elderly care service is higher than the overall evaluation, and the scores are 68.18 and 68.15 respectively. The satisfaction of the middle-aged group to the elderly care service is lower than the overall evaluation, which is 67.20, as shown in Figure 3.

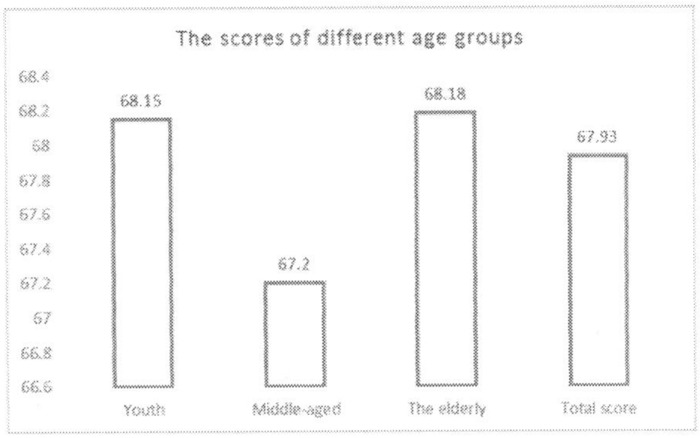

Fig. 3. Score of elderly care service satisfaction among different age groups in 2019.

3.2 Quality of elderly care service in different gender groups

According to the monitoring data, among the respondents who are willing to reveal their gender in the 2019 survey on the social satisfaction of the elderly care service in Liaoning Province, the number of men is 1,870, accounting for 51.05%, and that of women is 1,793, accounting for 48.95%, as shown in Table 2.

Table 2. The sex proportion of the respondents in the elderly care service quality satisfaction in 2019.

Gender	Gender count	% GT of Gender count
male	1870	51.05%
female	1793	48.95%

The monitoring results show that among the respondents who reveal their gender, female's satisfaction with elderly care service is higher than male's, among which female's satisfaction score is higher than the province's overall score of 68.08, and male's satisfaction is equal to the province's overall score of 67.93, as shown in Figure 4.

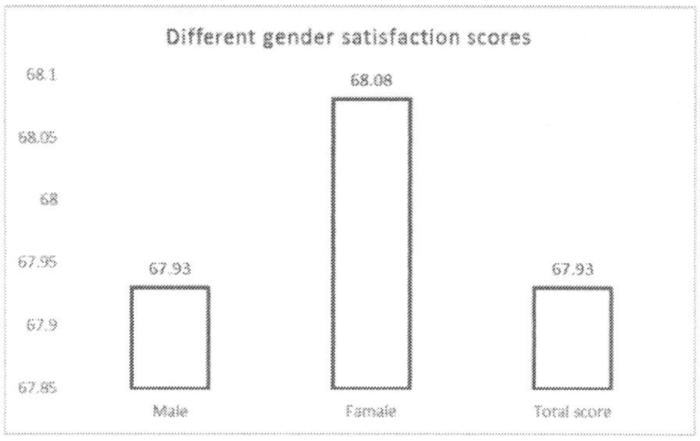

Fig. 4. Different Gender Public Satisfaction with Elderly Care Services.

3.3 The elderly care service quality for groups with different educational levels

According to monitoring data, among the respondents in the social satisfaction survey on the elderly care service quality in Liaoning Province in 2019, there are 158 people with primary school education, accounting for 4.30%; 588 people with junior high school education, accounting for 16.01%; 915 people with senior high school / secondary school / technical school education, accounting for 24.92%; 1,211 people with junior college education, accounting for 32.98%, ranking first; 738 people with undergraduate education, accounting for 20.10%; 62 people with master's degree or above, accounting for 1.69%, ranking last, as shown in Figure 5.

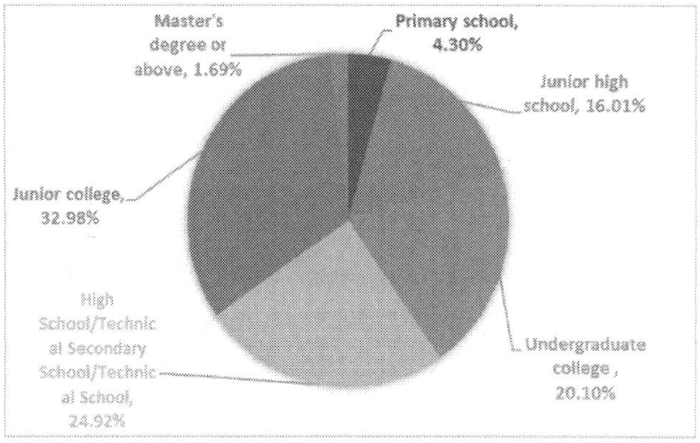

Fig. 5. The proportion of education level.

The satisfaction of the elderly care service is different among the groups with different education levels. With the increase of the group education, the satisfaction of the elderly care service is on the rise, as shown in Figure 6.

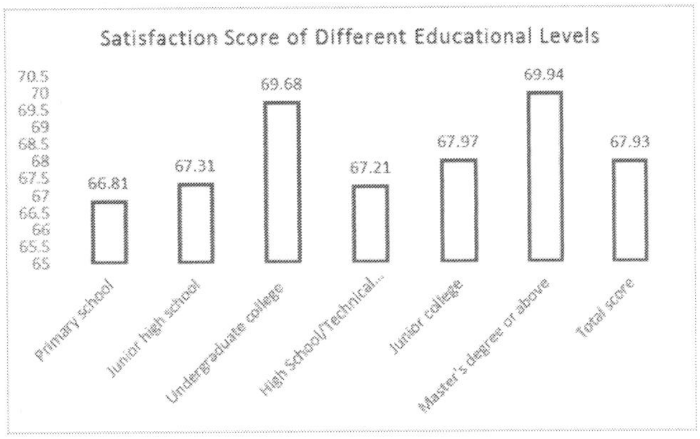

Fig. 6. Public satisfaction with elderly care services with different degrees.

3.4 Quality of service for the aged in different income groups

According to the statistics, among the urban and rural respondents who are willing to reveal their income in the 2019 social satisfaction survey on the elderly care service quality in Liaoning Province, the number of urban residents whose income is less than 2,000 yuan is 142, accounting for 4.68%; the number of urban residents whose income is between 2,001 and 4,000 yuan is 1,311, accounting for 43.17%; the number of urban residents whose income is between 4,001 and 6,000 yuan is 1,028, accounting for 33.85%; the number of urban residents whose income is between 6,001 and 10,000 yuan is 395, accounting for 13.01%; the number of urban residents whose income is between 10,001 and 20,000 yuan is 142, accounting for 4.68%; the number of urban residents whose income is 20,001 and above is 19, accounting for 0.63%.

The number of rural residents with income less than 1,000 yuan is 27, accounting for 4.35%; the number of rural residents whose income is between 1,001 and 2,000 yuan is 74, accounting for 11.92%; the number of rural residents whose income is between 2,001 and 4,000 yuan is 343, accounting for 55.23%; the number of rural residents whose income is between 4,001 and 6,000 yuan is 126, accounting for 20.29%; the number of rural residents whose income is between 6,001 and 10,000 yuan is 37, accounting for 5.96%; the number of rural residents whose income is 10,000 yuan and above is 14, accounting for 2.25%. The figure is shown in Figure 7, Figure 8.

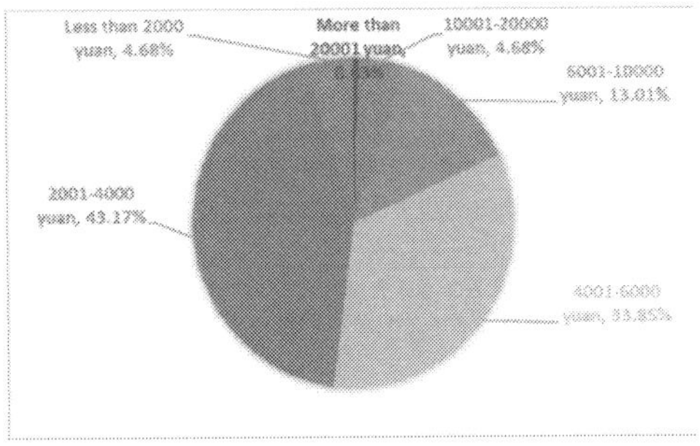

Fig.7. Urban Residents.

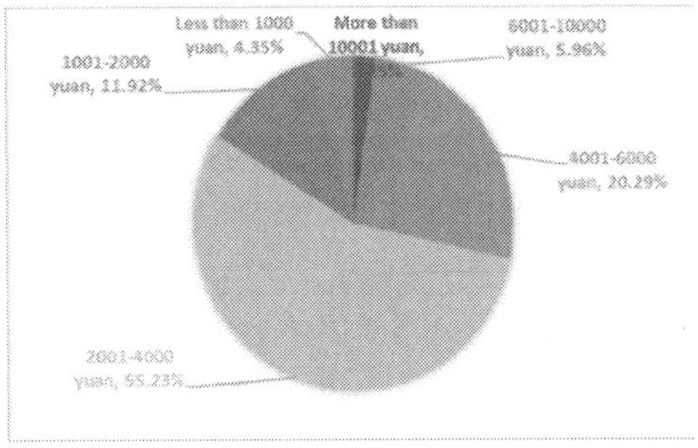

Fig. 8. Rural Residents.

The monitoring results show that the satisfaction of urban residents with the quality of elderly care service increases with the increase of average monthly family income, and the average is slightly lower than that of rural residents, indicating that the rate of improvement of elderly care service in rural areas is higher than that in urban areas.

Specific to each family average monthly income group, the group with family average monthly income between 10,001 yuan and 20,000 yuan has the highest social satisfaction evaluation, with the score of 69.90; the group with family average monthly income above 20,000 yuan has the lower social satisfaction evaluation, with the score of 66.57.

The rural residents' social satisfaction evaluation of the elderly care service quality basically decreases with the increase of the average monthly income of the family. Specifically, the social satisfaction evaluation of the people whose family average monthly income is less than 1,000 is higher, with the score of 71.57; the social satisfaction evaluation of the people whose family average monthly income between 4,001 yuan and 6,000 yuan is lower, with the score of 67.55. The results are shown in Figures 9 and 10.

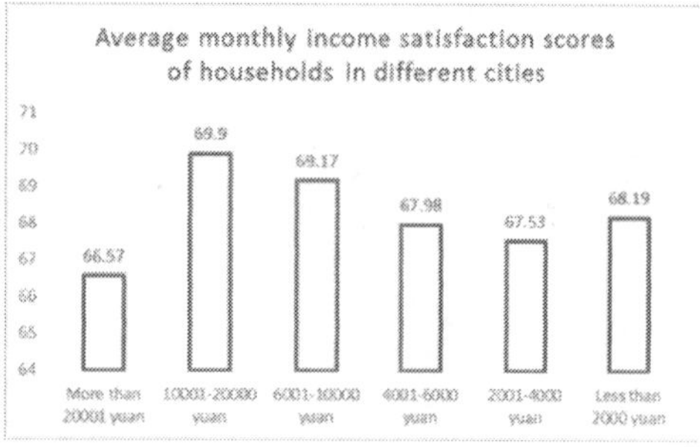

Fig.9. Income of Different Cities.

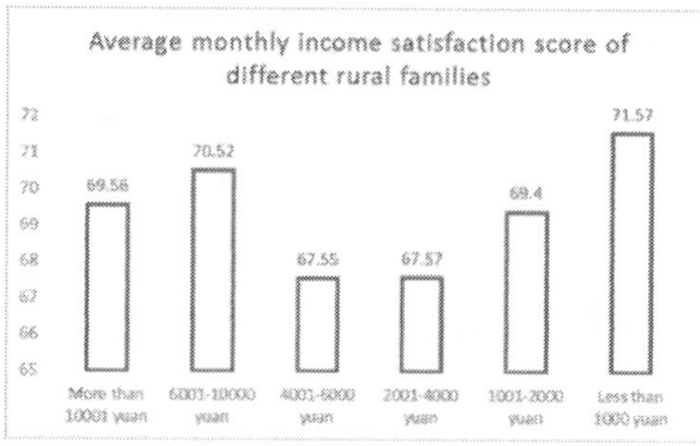

Fig. 10. Income of Different Rural.

4 Conclusion and suggestions

Taking the monitoring data of the elderly care service quality in Liaoning Province as an example, this paper analyzes the quality satisfaction of different groups in six fields: public elderly care service, community elderly care service, home-based elderly care service, the resettlement of the elderly, the preferential policies and services for the elderly, the public to the elderly care service and so on, and draws the following conclusions:

(1) There is little difference in the satisfaction of the elderly care service among the different age groups, and the satisfaction of the elderly care service is higher among the elderly group and the young group.

(2) Female's satisfaction with elderly care service is higher than male's.

(3) There is a great difference in the satisfaction of the elderly care service among the different education level groups, and the satisfaction of the elderly care service basically increases with the increase of the group education.

(4) The rural residents' social satisfaction evaluation of the elderly care service quality basically decreases with the increase of the average monthly income of the family. The urban residents' satisfaction evaluation of the elderly care service quality basically increases with the increase of the average monthly income of the family, and the average rate is slightly lower than that of the rural residents. The improvement rate of the elderly care service in rural areas is higher than that of the city.

In view of the above analysis results, it is suggested that according to the satisfaction of different groups, the government should carry out the targeted bench marking of the elderly care service quality, realize the precise connection between the demand and the supply of elderly care service quality, and improve the public's satisfaction with the elderly care service quality in a targeted manner so as to improve the whole level of the elderly care service quality.

References

1. D L Wang, X Z Song. China Administration. J. (6), 72-77(2009)
2. X G Luo, Hongyan Zhang. Administrative Forum. J. (3), 35-37(2008)
3. G F Cui. Hubei Social Sciences. J. (8), 37-38(2005)

4. W B Li, L H Lai. China Administration. J.(8), 53-57(2013)
5. E'e Zhang , J Zhang. Agricultural Science Research. J.(3), 92-96(2017)

Research on the Relationship between Psychological Safety and Individual Job Burnout

Xueyang Qi *and Yu* Wen*

College of Information and communication, National University of Defence Technology, Xi'an, China

Abstract. The development of globalization and the acceleration of informatization have made the competition between enterprises more and more fierce, and organizations must respond to new opportunities and challenges through organizational change. Organizational change also makes employees' working environment more and more complex, occupational uncertainties and risks are increasing, which in turn brings employees psychological insecurity and job burnout. In this paper, through combing the related research of psychological safety and job burnout, the negative relationship between psychological safety and individual job burnout is verified by empirical analysis, and according to the conclusion, relevant suggestions are proposed to the organization management, trying to help the organization take effective measures to satisfy employees' psychological safety needs and reduce the phenomenon of job burnout.
Keywords: Psychological safety, Job burnout, Individual.

1 Introduction

As a hot topic in organizational behavior, the emergence of the concept of psychological safety originates from the study of organizational change. Schein and Bennis [1] illustrated that it is necessary for organizations to pay attention to subordinates' psychological safety needs in the context of organizational change. In face of constant changes in the internal and external environment and the reintegration and alteration of resources, organizations must respond to opportunities and challenges, and win long-term survival and development through organizational changes, such as organizational structure adjustment, personnel restructuring, and strategic innovation. However, organizational change may induce organizational members' concerns, fears, and insecurity, which probably lead to increased turnover rate, slack in work, and reduced work performance and so on. Therefore, in order to help the members of the organization to overcome the uncertainty and insecurity brought about by organizational change from the psychological level, it is especially important to meet their psychological safety needs and strengthen the sense of organizational identification and trust [2]. Psychological safety not only reflects the safety needs of organizational members in the context of organizational change, but also is closely related to organizational members' work input. It can stimulate individual working enthusiasm and improve work performance, thus improving the core competitiveness of the organization.

* Corresponding author: xueyang_unimelb@163.com

Conversely, it will cause slack in work and occupational burnout if psychological safety is lacking. Therefore, it is of the great theoretical and practical significance to explore the relationship between psychological safety and individual job burnout.

Looking from the existing literature, scholars at home and abroad have already conducted adequate research on psychological safety. There is more literature about psychological safety in foreign countries, and the time span is wider. The main research angles can be summarized into three categories: First, the research on the concept of psychological safety. Foreign scholars divide the concept of psychological safety into three levels: individual, group and organization, and define them differently. The second is research on psychological safety measurement. Aiming at the different definitions of psychological safety at the individual, group and organizational levels, foreign scholars use different measurement scales when measuring psychological safety at different levels [3]. At the individual level, most of the psychological safety measurements use the psychological safety scale compiled by Tynan [4], which includes the two dimensions of self-psychological safety and others. At the group level, the team psychological safety scale developed by Edmondson [2] is also the classic and most cited scale. Brown and Leigh [5] divided psychological safety into three dimensions of supportive management, role clarity and self-expression from the organizational level for measurement. The third is the study of influencing factors of psychological safety, mainly from the aspects of individual characteristics, quality of interpersonal relationships, group activities and structures, leadership traits, and other situational factors to analyze how to affect the psychological safety of subordinates [6].

At present, domestic scholars mainly focus on the study of the impact and function mechanism of psychological safety and discuss the impact of psychological safety on organizational members' learning behavior, work performance, innovation behavior, work behavior, advocacy behavior and other aspects. For example, Chen, Zhao and Jiang [7] used empirical research methods to deeply analyze the relationship between psychological safety and the nine sub-capabilities of team learning, and verified that psychological safety can affect team performance through the intermediary role of the overall learning ability of the team ; Zhang [8] studied the relationship between the three variables of psychological safety, knowledge self-confidence and team members' willingness to share knowledge; Shi [9] deeply explored the mediating effect of psychological safety on organizational trust and employee innovation behavior; Yuan and Liu [10] used 460 corporate employees as a sample to discuss the moderating role of psychological safety in the relationship between organizational ethics and employee voice behavior. Most of the existing research focuses on the behavioral results produced by psychological safety, while research on the effects of psychological safety on work attitudes is still deficient. Although variables such as work engagement, work involvement, and work input have been confirmed to be affected by psychological safety, the current research mainly focuses on the impact of psychological safety on positive outcome variables, and is lacking in the study of the impact of psychological safety on negative outcome variables. Based on this, this article takes the negative result variable of job burnout as the research object, and explores the relationship between psychological safety and job burnout from the concepts of psychological safety and job burnout. It is hoped that through the research, the impact and function mechanism of psychological safety on individual job burnout will be revealed.

2 Literature review

2.1 Psychological safety

There are many definitions of the concept of psychological safety in academia. Overall, according to different levels of attention, they can be divided into three levels: individual, group and organization. As shown in Figure 1.

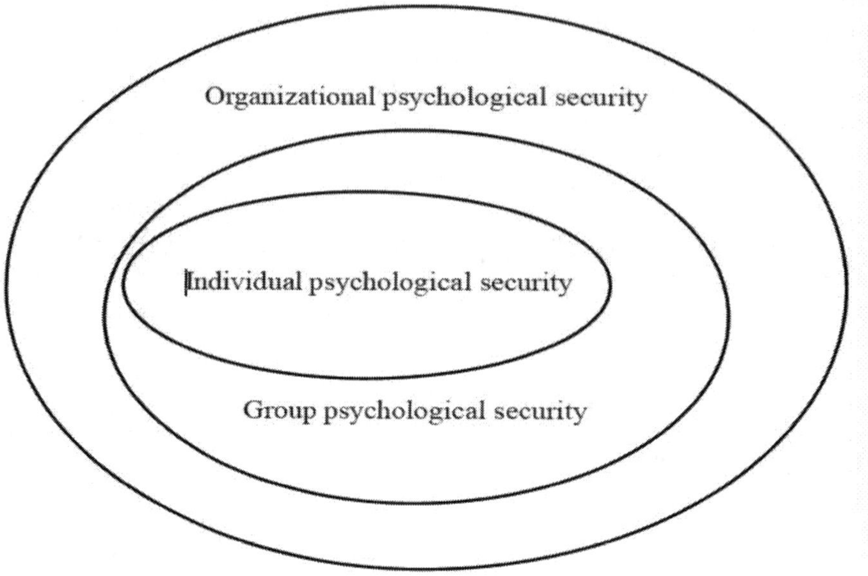

Fig. 1. Multi-layered concept of psychological security.

Kahn [11] first proposed the concept of individual psychological safety in the study of the impact of employee psychological safety on work engagement. He believes that individual psychological safety is an individual characteristic, which reflects the self-perception of the internal psychological state of the members of the organization, and defines psychological safety as: the organization members can express and show themselves without worrying about the negative impact of the behaviors on their personal image, status and work, which is the most recognized definition in the academic circles when studying individual psychological safety. Essentially, individual psychological security is the individual's intrinsic motivation and the mental state of shaping the individual's psychological role. When the members of the organization have a strong sense of psychological security, it will increase work input and participation; on the contrary, when the sense of psychological security decreases, job burnout increases.

The concept of group psychological safety originated from the impact and changes brought about by economic development on the traditional organizational structure. With the increasingly fierce organizational competition, the traditional organizational structure can not adapt to the new situation and new environment, the team as a more efficient and flexible new organizational model has been recognized by more and more managers, so the concept of team psychological safety comes into being. Edmondson [2] first proposed the concept of group psychological safety from the perspective of team characteristics in the study of team learning behavior, and defined it as a shared belief that the team ensures that "interpersonal risk is safe." Psychological safety at the group level describes the overall characteristics of the group, not the characteristics of individual group members. It is closely related to interpersonal trust between groups, but it also transcends interpersonal trust.

Organizational psychological security is represented by the shared beliefs of the entire organization, which belongs to the characteristics of organizational atmosphere. Brown and Leigh [5], when studying the relationship between psychological atmosphere, job involvement and job performance, defined organizational level psychological safety as organizational members' perception of organizational atmosphere and organizational environment, including perceived supportive management, role clarity and organizational characteristics that allow self-expression. When the organization adopts more supportive management, gives employees clear role expectations, and allows employees to express their views freely, members of the organization will have a higher sense of psychological security. Baer and Frese [12] believe that organizational psychological safety refers to the formal or informal organizational norms and procedures that support the open interaction of members of the organization.

Since this article studies the relationship between psychological safety and individual job burnout, the concept of individual layer psychological safety is selected.

2.2 Job burnout

Job burnout was originally discovered by psychologist Herbert Freuden-berger in the 1860s when his colleagues showed a series of negative emotional and behavioral performances at work, and the professional term "burnout" was first proposed in 1974. Job burnout is used to describe the negative states such as cynicism, depression, depression, and boredom that employees show at work [13]. Maslach and Leiter [14] divide the symptoms of job burnout into three dimensions: emotional exhaustion, depersonalization, and reduced sense of accomplishment. Among them, emotional exhaustion refers to feelings of excessive consumption and commitment; depersonalization refers to apathy and negative attitudes or emotions towards others at work; reduced sense of accomplishment refers to dissatisfaction with one's own work performance and ability.

3 Research methods

3.1 Research Hypothesis

At present, domestic scholars have relatively little research on the relationship between psychological safety and job burnout. After searching, Liang [15] used 558 doctors, nurses and technicians as the research object to discuss the psychological mechanism of medical workers 'job burnout. Through empirical research, it was found that medical workers generally have job burnout, and improving employees' psychological security can effectively reduce job burnout. From the perspective of job insecurity brought by layoffs, Chen [16] established a new model of the relationship between job insecurity and job burnout of the new generation of employees, and concluded that sense of security is positively related to job burnout. Based on the above discussion, this article proposes the following assumptions:

H1: psychological safety has a negative effect on job burnout ;

H1-1: psychological safety has a negative effect on emotional exhaustion ;

H1-2: psychological safety has a negative effect on depersonalization ;

H1-3: psychological safety has a negative effect on the reduction of sense of accomplishment ;

3.2 Research object

The samples in this study are teachers, corporate employees and government staff in Xi'an, Shanghai, Suzhou, Beijing and other places. In order to improve the reliability and effectiveness of the questionnaire survey, the research objects involve various industries as much as possible. The questionnaire is distributed mainly through the app "Questionnaire Star", and then the questionnaire is recovered in time through the app. The questionnaire survey and collection work lasted for one month. A total of 500 questionnaires were distributed and 420 valid questionnaires were recovered. Among them, in terms of gender, men account for 41% and women for 59%. For age, 25 years and younger account for 10%, 26-30 years for 35%, 31-35 years for 27%, 36-40 years for 22%, 41% and below accounted for 6%. In terms of education level, 36% undergraduate, 61% master and 3% doctor. In terms of job nature, state-owned enterprises accounted for 23%, private enterprises accounted for 35%, public institutions and government departments accounted for 42%. In terms of working years, 31% within 5 years, 43% within 5-10 years, and 26% over 10 years.

3.3 Research tools

Individual level psychological safety scale: using the Likert 9-point interpersonal interaction psychological safety scale compiled by Tynan [4]. The scale includes two dimensions of self-psychological safety and the psychological safety of others, with a total of 12 items. The self-psychological safety scale includes 7 items, and the others' psychological safety scale includes 5 items.

Work Burnout Scale: The work burnout scale developed by Maslach has a total of 12 items, including three dimensions: emotional exhaustion, depersonalization, and reduced sense of accomplishment. The scale uses Likert 7-point scoring method, the higher the score, the more serious the job burnout.

4 Conclusion

This article uses Amos17.0 software to perform reliability and validity analysis, descriptive statistical analysis, differential impact analysis and regression analysis on the scale to verify the relationship between psychological safety and individual job burnout. The specific verification situation is shown in the Table 1.

This study concludes that psychological safety is significantly negatively correlated with job burnout. The higher the psychological safety score, the lower the job burnout, which is consistent with previous research results. Therefore, psychological safety, as a source of occupational stress that cannot be ignored, will have an important impact on the physical and mental health and work motivation of the members of the organization. Only when employees have a strong sense of psychological security, can feel the stability guarantee provided by the organization, reasonable promotion opportunities and development space, will they see the future development prospects, and thus maintain their passion for work and are willing to devote themselves to work. Otherwise, job burnout will occur. It is suggested that the organization should establish a perfect and reasonable performance evaluation system, evaluation system, training system and effective communication channels to enhance employees' psychological security, thereby decreasing the phenomenon of job burnout.

Table 1. Summary table of verification of various theoretical assumptions.

No.	Assumption Content	Verification Results
H1	Psychological safety has a negative effect on job burnout	Support
H1-1	Psychological safety and emotional exhaustion have a negative impact	Support
H1-2	Psychological safety and depersonalization have a negative impact	Support
H1-3	Psychological safety and reduced sense of accomplishment have negative effects.	Support

References

1. E H Schein, W Bennis. Personal and organizational change via group methods. New York: Wiley (1965)
2. A Edmondson. *Psychological Safety and Learning Behavior in Work Teams [J]. SAGE Publications*, **44** (2) (1999)
3. Y C Zou, T B Yin. *A review of psychological security research from a multi-level perspective* [J]. China Human Resources Development (2017)
4. R Tynan. *The effects of threat sensitivity and face giving on dyadic psychological safety and upward communication*. Journal of Applied Social Psychology, **35** (2), 223–247 (2005)
5. S P Brown, T W Leigh, *A new look at psychological climate and its relationship to job involvement, effort, and performance*. Journal of Applied Psychology, **81** (4), 358–368, (1996)
6. B Ling, J Y Duan, Y L Zhu. *Psychological safety in the workplace: conceptual design, influencing factors and results* [J]. Advances in Psychological Science, **18** (10), 1580-1589 (2010)
7. G Q Chen, H Q Zhao, L J. *An empirical study on the relationship between team psychological safety, team learning ability and team performance* [J]. Science Research, **26** (06), 1283-1292 (2008)
8. T Zhang. *Research on the influence of psychological safety and knowledge self-confidence on the willingness of team members to share knowledge* [D]. Central China Normal University (2014)
9. T Shi. *Research on the impact of intra-organizational trust on employee innovation behavior* [D]. Nanjing University of Technology (2017)
10. L Yuan, Z Y Liu. *Research on the relationship between organizational ethical atmosphere and employee behaviors from the perspective of psychological security* [J]. China Human Resources Development (2016)
11. A. Kahn William. *Psychological Conditions of Personal Engagement and Disengagement at Work* [J]. Academy of Management, **33** (4) (1990)
12. M Baer, M Frese. *Innovation is not enough: climates for initiative and psychological safety, process innovations, and firm performance* [J]. John Wiley & Sons, Ltd., **24** (1) (2003)
13. C D Shepherd, A Tashchian, R E Ridnour. *An Investigation of the Job Burnout Syndrome in Personal Selling* [J]. Routledge, **31** (4) (2011)

14. Y L Liang. *The effect of inclusive leadership and psychological security on medical staff 's job burnout* [D]. Yunnan Normal University (2019)
15. M J Chen. *The impact of job insecurity of new generation employees on occupational burnout—the moderating role of work pleasure* [D]. Nanjing Normal University (2015)

A Study on the Impact of Consumer Ethics on Apparel Purchasing Behavior

Hongshan ZHAO*

Beijing Institute of Fashion Technology, Beijing, China

Abstract. As the Chinese economy grows fast and the Chinese residents' spending power increases, "ethical consumers" have been lavished with increasing attention. In fact, consumption has become a decision–making process closely tied to ethics or morality. This article describes consumers' basic cognition of environmental issues and ethical apparel in a quantitative manner. The research covers mainly the following topics: consumers' understanding and recognition of ethical apparels, factors bearing on the behavior of purchasing ethical apparels, surveys of the intentions of consumers to purchase ethical apparels as well as the personal information of survey respondents. The purpose is to provide forward-looking bases and reference for apparel enterprises to plan ethical marketing.

Keywords: Consumption, Consumer Ethics, Consumer Ethical Behavior, Apparel Ethical Marketing.

1 Introduction

A study of the UK consumers and producers indicates that a growing number of people begin to realize that the production and consumption of apparel may deal a negative impact on ecological environment. As a result, the textile and apparel industries have seen increasingly prevalent ethical consumption. In the business history of the US, there were exemplary enterprises that have both seen success and observed ethics. Warren Buffett mentioned in a speech for students of the University of Columbia Business School that "In this country, those complying with ethical norms can harvest great success. Ethics will not burden you. Sometimes, it is neutral; sometimes, it can even lend you a helping hand. But it is by no means a burden to you." From a long-term perspective, the core of business is none other than relations and trust, so an ethical way of doing business is no doubt the ultimate way to success.

As a special consumer segment, "ethical consumers" have drawn increasing attention. Because consumers of this type have a deep sense of responsibility with respect to environment and society and attempt to articulate their appeals on ethical values through ethical purchasing or resistant behavior, consumption for them has almost become a decision-making process closely tied to ethics or morality. The attention to and belief in ethical issues can effectively shape consumers' purchasing behavior. This article aims to investigate and

* Corresponding author: sxyzhs@bift.edu.cn

analyze the impact of consumer ethics on apparel purchasing decisions. Its conclusions may provide some reference for enterprises to undertake ethical fashion marketing.

2 Connotation of the view of consumer ethics

The view of consumer ethics is a compound notion. It has always been the most crucial dominant part of consumer attitude. It is a result of mutual assimilation between the notion of consumer ethics and the notion of consumer attitude. The view of consumer ethics refers to ideological concepts that have taken shape by making certain moral evaluation of various consumption styles and behaviors that occur in everyday life and that are used to guide specific consumption behaviors.

The view of consumer ethics is based on the consumer ethics, and is in itself a subjective consciousness within the notion of consumer ethics. The importance of consumer ethics view finds expression in the fact that it directly bears on the consumption style and behavior of an individual or a particular consumer group. In fact, the consumer ethics view always uses ethical or moral standards to understand or evaluate consumer behaviors. To put it simple, if a consumption concept is of ethical nature, it is then a view of consumer ethics, for it constantly focuses on whether a consumption behavior is justifiable or not and whether it enables people to live a happy life or not.

3 Theoretical foundation

In the consumer behavior research area, an absolute majority of scholars believe that intention is the most important indicator of behavior. Therefore, to correctly understand the ethical consumption behavior of a consumer with regard to apparel, we must first of all examine his/her ethical purchasing intention. Intention means the degree to which a consumer attempts to perform a particular behavior [1] or the level of effort he/she is willing to make in order to achieve a particular behavior [2]. For researching purchasing intention, the theory of planned behavior is considered to be one of the most powerful models. Many empirical studies have demonstrated the capability of the planned behavior theory to explain various behavioral intentions. In reality, however, ethical purchasing decisions of a consumer are quite complicated. As Boulstridge and Carrigan [3] see it, consumers may care about the ethical behavior of an enterprise, but do not necessarily have the intention to do ethical purchasing, nor do they necessarily support ethical products or resist non-ethical products. This article examines the factors that affect consumers' ethical purchasing behaviors based on the planned behavior theory.

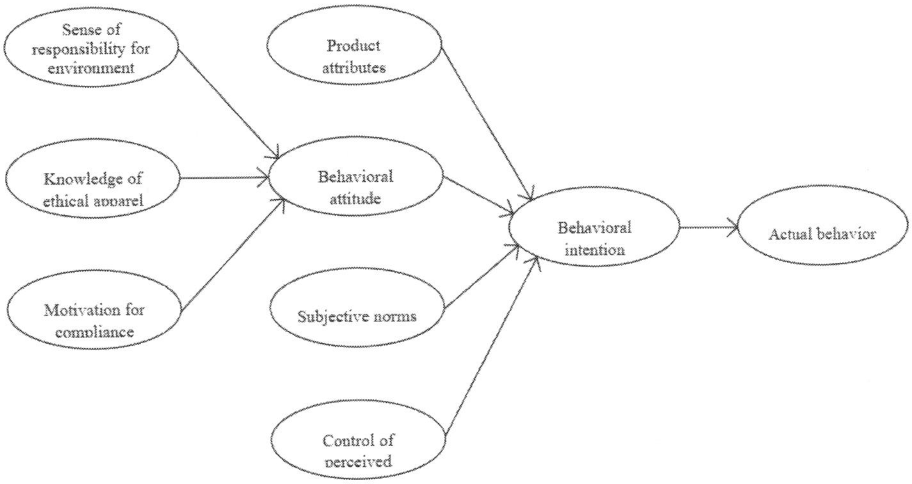

Fig. 1. Research Model Adopted by This Article.

Based on previous studies, this article explores from the psychological perspective the relationship between product attributes, behavioral attitude, subjective norms and control of perceived behavior on one side and the intention to purchase ethical apparel on the other side. In particular, this article holds that we should examine the behavioral attitude of a consumer in terms of the sense of responsibility for environment, the knowledge of apparel ethics, and the motivation for compliance. The research model is charted in Figure 1. On this basis, this article puts forward the following hypotheses:

H1: Product attributes have significant positive effect on the ethnical purchasing intention of a consumer.

H2: Behavioral attitude has significant positive effect on the ethnical purchasing intention of a consumer.

H3: Subjective norms have significant positive effect on the ethnical purchasing intention of a consumer.

H4: Control of perceived behavior has significant positive effect on the ethnical purchasing intention of a consumer.

This article selects some factors that have relatively greater influence on the purchasing of ethical apparel. Such factors primarily include the following:

3.1 Product attributes

The attributes of apparel products include prices, colors, styles, fabric and workmanship, among others. Moreover, the attributes of apparel products are consistent with purchasing intention, so much so that the good quality and reasonable price of an apparel product would positively bolster the purchasing intention of a consumer.

3.2 Behavioral attitude

Behavioral attitude represents a person's willingness to perform a particular behavior in a given context, as well as his/her assessment, whether positive or negative [4], of the behavior in its entirety. Based on previous studies, this article examines the impact of three factors, namely, the sense of responsibility for environment, the knowledge of apparel ethics and the motivation for compliance, on the behavioral attitude of consumers. The first two factors are relatively easy to understand, and the third, i.e. the motivation for compliance, is the willingness of an individual to perform ethical consumption behavior by complying with the intention of other persons important to him/her. Chinese consumers tend to comply with social norms or the expectations of others, the collective or a group. Those who motivate consumers to comply with include not only family members, friends and colleagues, but also government, media and consumer protection groups.

3.3 Subjective norms

Subjective norms mainly refer to the impact of external social environment on the behavior of an individual; they are presented as the social pressure perceived by an individual as a result of performing or not performing a particular behavior [5]. The pressure arises mainly when other persons important to him/her, like his/her family members and friends, think it is desirable or undesirable for him/her to perform a given behavior [6]. When a person perceives that people in his/her surroundings will support or affirm his/her behavior and he/she takes a positive attitude toward the behavior, he/she will have a very strong intention to perform that behavior.

3.4 Control of perceived behavior

The control of perceived behavior is an important variable in the theory of planned behavior. It is an individual's perception of the various obstacles he/she may encounter when performing a particular behavior. In other words, it is an individual's judgment of the level of difficulty when he/she performs a particular behavior according to his/her previous experience or expectations. When a person believes that he/she has relatively ample resources and opportunities (e.g. money, skills, help, etc.) and expects relatively little difficulties, the control of perceived behavior will be strong. To the contrary, if a person feels helpless in purchasing ethical apparel, his/her behavioral intention will be checked to a certain degree.

4 Empirical study

This article adopts the empirical analysis method. The empirical analysis consists mainly of the following steps: questionnaire design, testing, sampling and data collection, and data analysis.

The questionnaire for this study covers four aspects as follows:

(1) consumers' basic understanding of environmental issues and ethical apparel

We will try to understand consumers' understanding of environmental issues and ethical apparel by examining their knowledge of low carbon emission, environmental protection and ethical apparel and their recognition of ethical apparel.

(2) factors influencing consumers' ethical apparel purchasing behavior

We will examine the factors that influence consumers' ethical apparel purchasing behavior by looking at such elements as prices, colors and styles, fabric and workmanship, motivation for compliance, media and information.

(3) investigation of consumers' intention to purchase ethical apparel

We will study consumers' intention to purchase ethical apparel from such perspectives as prices, whether or not they will purchase ethical apparel in the future, and premium rate.

(4) personal information of survey respondents

Such information includes gender, age, occupation, employment, education, monthly income, and other basics. Respondents are required to answer every question and evaluate

 https://doi.org/10.1051/shsconf/20219602005

the degrees to which they agree with statements in the questionnaire. The questionnaire adopts a 7-point Likert scale, with 1 representing "completely disagree" and 7 "completely agree".

The sample survey was mainly conducted in Beijing. The sample was divided into two parts, namely the part for university students and the part for non-university students. The sampling was undertaken at the same time. A total of 260 questionnaires were distributed, with 245 retrieved. After a consistency check, 45 questionnaires were screened out. So there were a total of 200 effective questionnaires, with the recovery rate standing at 94.23% and the effective rate at 76.92%.

5 Descriptive statistical analyses

We have made descriptive statistical analyses on the samples surveyed. The results of such analyses are shown in Table 1.

Table 1. Descriptions of the Samples Surveyed.

Group variable		Frequency	Percentage	Effective percentage	Accumulated percentage
Gender	male	93	46.5	46.5	46.5
	female	107	53.5	53.5	100.0
Monthly income	Less than 1,000	16	8	8	8
	1,000-3,000	41	20.5	20.5	28.5
	3,000-5,000	48	24.0	24.0	52.5
	5,000-8,000	33	16.5	16.5	69.0
	8,000-10,000	56	28.0	28.0	97.0
	More than 10,000	6	3.0	3.0	100.0
Education	Secondary technical schooling or below	6	3.0	3.0	3.0
	Junior college	10	5.0	5.0	8.0
	Undergraduate	147	73.5	73.5	81.5
	Master's degree or above	37	18.5	18.5	100.0
Age	Younger than 20	8	4.0	4.0	4.0
	20-30	89	44.5	44.5	48.5
	30-40	59	29.5	29.5	78.0
	40-50	30	15.0	15.0	93.0
	Older than 50	14	7.0	7.0	100.0
Occupation	Government employees	20	10.0	10.0	10.0
	Company employees	30	15.0	15.0	25.0
	Students	100	50.0	50.0	75.0
	Education, research and medical workers	19	9.5	9.5	84.5
	Self-employed individuals	24	12.0	12.0	96.5
	Others	7	3.5	3.5	100.0
Effective sample size	200				

This article uses SPSS19.0 to calculate the maximum value, minimum value, mean value and standard deviation of all variables observed. The results of calculation are shown in

Tables 2 to 6. The mean values of all observed variables for knowledge of apparel ethics are generally on the low side, with the minimum value at 2.58 (the second indicator), indicating that consumers know relatively little about ethical apparel. The mean values of all other variables are on the upper medium level. The standard deviations for all variables stand above 1, indicating a relatively great difference in the choices of respondents in answering the survey questions.

Table 2. Descriptive Statistical Analyses on the Latent Variable "Product Attributes".

	Observed variable	Sample size	Minimum value	Maximum value	Mean value	Standard deviation
Latent variable	A3.1	200	1	6	4.89	0.955
	A3.2	200	1	7	4.88	1.110
	A3.3	200	1	7	4.96	0.926
Effective sample size		200				

Table 3. Descriptive Statistical Analyses on the Latent Variable "Behavioral Attitude"

Latent variable	Latent sub-variable	Observed variable	Sample size	Minimum value	Maximum value	Mean value	Standard deviation
Behavioral attitude	Sense of responsibility for environment	A1.1	200	2	7	4.35	1.132
		A1.2	200	1	7	4.51	0.997
		A1.3	200	1	7	4.56	0.975
		A1.4	200	2	6	4.62	0.825
	Knowledge of apparel ethics	A2.1	200	2	6	3.01	1.132
		A2.2	200	2	7	2.58	0.887
		A2.3	200	1	6	3.50	1.032
		A2.4	200	1	6	4.56	1.270
	Motivation for compliance	A5.1	200	1	7	4.69	1.028
		A5.2	200	1	7	4.69	1.171
Effective sample size			200				

Table 4. Descriptive Statistical Analyses on the Latent Variable "Subjective Norms".

Latent variable	Observed variable	Sample size	Minimum value	Maximum value	Mean value	Standard deviation
Subjective norms	A4.1	200	1	7	4.80	1.144
	A4.2	200	2	6	4.46	0.955
	A4.3	200	1	7	4.53	1.143
	A4.4	200	1	7	4.75	0.934
Effective sample size		200				

Table 5. Descriptive Statistical Analyses on the Latent Variable "Control of Perceived Behavior".

Latent variable	Observed variable	Sample size	Minimum value	Maximum value	Mean value	Standard deviation
Control of perceived behavior	A6.1	200	1	7	5.09	0.881
	A6.2	200	1	7	5.06	0.897
	A6.3	200	3	6	4.82	0.837
Effective sample size		200				

Table 6. Descriptive Statistical Analyses on the Latent Variable "Behavioral Intention".

Latent variable	Observed variable	Sample size	Minimum value	Maximum value	Mean value	Standard deviation
Behavioral intention	A7.1	200	3	7	5.31	0.741
	A7.2	200	2	6	4.88	0.940
	A7.3	200	2	7	4.87	0.791
Effective sample size		200				

The premium of a product can be seen as the difference between the price of the product and the average price of similar products. If the price of ethical apparel is set too high, consumers may not be willing to purchase ethical apparel at the cost of part of their personal interest. Therefore, the premium rate of ethical apparel must be controlled within a reasonable range to make it more easily accepted by consumers. The premium rates acceptable to consumers as shown by studies in this article are presented in Table 7. It can be seen that 89.5% of the respondents accept a premium rate of lower than 15% for ethical apparel.

Table 7. Premium Rates Acceptable to Consumers.

Group variable	Frequency	Proportion of consumers who accept	Effective percentage	Accumulated percentage
Lower than 5%	25	12.5	12.5	12.5
5-10%	36	18.0	18.0	30.5
10-15%	118	59.0	59.0	89.5
15-20%	7	3.5	3.5	93.0
20-25%	8	4.0	4.0	97.0
Higher than 25%	6	3.0	3.0	100.0
Total	200	100.0	100.0	

The times of ethical apparel purchasing done by consumers in the past are shown in Table 8. 24.0% of the consumers surveyed had never purchased ethical apparel. Even if they had done so, the frequency was extremely low.

Table 8. Ethical Apparel Purchasing History.

Group variable	Frequency	Percentage	Effective percentage	Accumulated percentage
Never	48	24.0	24.0	24.0
Once every 10 instances of apparel purchasing	78	39.0	39.0	63.0
2 to 3 times every 10 instances of apparel purchasing	51	25.5	25.5	88.5
4 to 5 times every 10 instances of apparel purchasing	15	7.5	7.5	96.0
6 to 7 times every 10 instances of apparel purchasing	4	2.0	2.0	98.0
8 to 9 times every 10 instances of apparel purchasing	1	0.5	0.5	98.5
Always	3	1.5	1.5	100.0
Total	200	100.0	100.0	

6 Reliability analyses

This article uses Cronbach's α as the standard to check the reliability of the questionnaire. The CITC must not be lower than 0.5; otherwise, the tested item concerned will be removed. This article conducts reliability analyses with respect to product attributes, behavioral attitude, subjective norms, control of perceived behavior, and behavioral intention in the original scale, with the results of analyses shown in Table 9. As shown in the table, the CITC value of A2.4 and A4.1 is 0.201 and 0.175 respectively, so the two items are deleted. The results also indicate that the Cronbach's α of various items is higher than the critical value of 0.6, fully indicating the relatively high reliability of the data for the variables studied. Moreover, the CITC values of all variables are greater than 0.4, indicating that all items are within the meaning of the variables and can truly describe the meaning of variables. Furthermore, deleting any item will not significantly raise the value of Cronbach's α, indicating that the scale adopted in this article has a relatively high level of internal consistency.

Table 9. Results of Scale Reliability Analyses.

Latent variable	Identifier variable	CITC (CITC)	Deleted α value under item	α value	α value
Product attributes (3 questions)	A3.1 (the degree to which prices sway my ethical apparel purchasing behavior)	0.730	0.841	0.83	0.81
	A3.2 (the degree to which colors sway my ethical apparel purchasing behavior)	0.651	0.739		
	A3.3 (the degree to which fabric and workmanship sway my ethical apparel purchasing behavior)	0.654	0.732		
Behavioral attitude (10 questions)	A1.1 (I am willing to sacrifice my personal interest for environmental protection)	0.753	0.753	0.78	0.76
	A1.2 (I will take the initiative to communicate environmental protection knowledge)	0.644	0.644		

	A1.3 (I will use degradable shopping bags when shopping)	0.520	0.520		
	A1.4 (I will consider whether the enterprise concerned has the sense of social responsibility when purchasing)	0.523	0.523		
	A2.1 (I know ethical apparel quite well)	0.705	0.705		
	A2.2 (I will check whether the product has environmental protection logos when purchasing)	0.562	0.562		
	A2.3 (I can correctly identify ethical apparel)	0.552	0.552		
	A2.4 (There are many ethical apparel options on the current market)	0.201	0.201		
	A5.1 (If people around me often buy ethical apparel, I will buy it, too)	0.533	0.533		
	A5.2(If government or media call on me to buy ethical apparel, I will)	0.591	0.591		
Subjective norms (4 questions)	A4.1 (Choosing non-ethical apparel will make me feel guilty)	0.175	0.175	0.79	0.73
	A4.2 (Purchasing ethical apparel gives me a sense of pride)	0.583	0.583		
	A4.3 (Purchasing ethical apparel can win respect from others)	0.726	0.726		
	A4.4 (I may choose to buy ethical products if I am willing to)	0.587	0.587		
Control of perceived behavior (3 questions)	A6.1 (I will buy it often if I have sufficient money, time or opportunity)	0.633	0.633	0.77	0.72
	A6.2 (I will buy ethical products if there are sufficient referential information)	0.621	0.621		
	A6.3 (I will buy ethical products if there are sufficient options of such products)	0.534	0.534		
Behavioral intention (3 questions)	A7.1 (I will buy ethical products first under the same prices)	0.665	0.665	0.81	0.80
	A7.2 (I will buy ethical products even if they are more expensive)	0.559	0.559		
	A7.3 (I will try buying ethical apparel after this survey)	0.647	0.647		

Note: As the CITC value for A2.4 (There are many ethical apparel options on the current market) and A4.1 (Choosing non-ethical apparel will make me feel guilty) stands at 0.201 and 0.175 respectively, the two items are therefore deleted.

7 Confirmatory factor analyses

Validity is the ability of a scale to measure the thing it is designed to measure [7]. This article chooses confirmatory factor analyses to test the validity of the model adopted in this article. Factor analyses are methods that explore whether there are potential factors that cannot be directly observed but may play a dominant role with regard to changes in the observable correlated variables. They are methods to look for potential factors that play dominant roles. Confirmatory factor analyses are mainly used to check the accuracy of testing models and whether the observed variables can fully represent latent variables, as well as the overall reliability and validity of the questionnaire.

We use statistical software to conduct factor analyses on the data collected. After strictly screening and checking the observed variables, we find that the extracted value of A5.2 stands at 0.5, which is relatively low and therefore cannot represent the data well. Moreover, the extracted value of items A3.3, A1.3 and A6.2 stands at 0.525, 0.525 and 0.512 respectively. They are all on the low side and therefore relatively poorly representative of the data concerned, so they are all deleted. Tables 10 to 13 present the results of post-deletion analyses. Generally speaking, when KMO is higher than 0.7, the data concerned are basically suitable for factor analyses. Table 10 presents the results of KMO and Bartlett's tests after data screening. It is clear that the result of the KMO test is 0.804, higher than the required 0.7, indicating that the data are highly suitable for factor analyses. The significance level for the Bartlett's test is 0.000, far lower than 0.05, also indicating the suitability of the data for factor analyses.

Table 10. KMO and Bartlett's Tests.

Kaiser-Meyer-Olkin measure of sampling adequacy	0.804
Chi-squared approximation for Bartlett's test	350.904
df	136
Sig.	0.000

Table 11. Common Factor Variance (Second Time).

Items	Initial	Extracted
A3.1	1.000	0.679
A3.2	1.000	0.682
A1.1	1.000	0.669
A1.2	1.000	0.689
A1.4	1.000	0.574
A2.1	1.000	0.720
A2.2	1.000	0.775
A2.3	1.000	0.659
A5.1	1.000	0.683
A4.2	1.000	0.533
A4.3	1.000	0.599
A4.4	1.000	0.634
A6.1	1.000	0.707
A6.3	1.000	0.698
A7.1	1.000	0.589
A7.2	1.000	0.660
		0.706
A7.3	1.000	

Table 12. Total Variance Explained.

Component	Initial eigenvalue			Extraction sums of squared loadings			Rotation sums of squared loadings		
	Total	% of variance	Accumulated %	Total	% of variance	Accumulated %	Tot-al	% of variance	Accumulated %
1	3.425	25.299	25.299	3.425	25.299	25.299	4.021	32.786	32.786
2	3.211	23.710	49.009	3.211	23.710	49.009	2.980	24.304	57.090
3	1.706	12.600	61.609	1.706	12.600	61.609	1.231	10.039	67.129
4	1.125	8.311	69.920	1.125	8.311	69.920	0.342	2.791	69.920
5	0.654	4.831	74.751						
6	0.531	3.922	78.673						
7	0.476	3.516	82.189						
8	0.476	3.516	85.705						
9	0.312	2.305	88.010						
10	0.260	1.921	89.931						
11	0.232	1.714	91.645						
12	0.216	1.595	93.240						
13	0.202	1.492	94.732						
14	0.202	1.492	96.224						
15	0.200	1.477	97.701						
16	0.200	1.477	99.178						
17	0.081	0.822	100.000						

Table 13. Rotated Component Matrix.

Item	Component 1	Component 2	Component 3	Component 4
A3.1	0.825	0.125	0.342	0.290
A3.2	0.067	0.234	0.378	0.694
A1.1	0.321	0.098	0.521	0.225
A1.2	0.635	0.257	0.136	0.289
A1.4	0.901	0.237	0.102	0.106
A2.1	0.034	0.432	0.531	0.369
A2.2	0.261	0.394	0.724	0.105
A2.3	0.098	0.278	0.038	0.844
A5.1	0.385	0.274	0.641	0.056
A4.2	0.202	0.156	0.373	0.742
A4.3	0.407	0.162	0.236	0.178
A4.4	0.359	0.270	0.503	0.314
A6.1	0.632	0.269	0.162	0.342
A6.3	0.544	0.170	0.210	0.219
A7.1	0.179	0.538	0.366	0.272
A7.2	0.369	0.491	0.274	0.215
A7.3	0.359	0.274	0.721	0.134

Table 12 lists the total variance contribution ratio for the factor analyses. It shows that four factors are extracted in the end and their variance contribution ratio to the entire variables stands at 25.299%, 23.710%, 12.600% and 8.311% respectively. The factors correspond to latent variables: behavioral attitude, subjective norms, product attributes, and control of perceived behavior respectively, with their total contribution ratio reaching 69.920%.

From the results of the foregoing confirmatory factor analyses, we arrive at the conclusion that the four hypotheses made by this study all hold water. Nevertheless, behavioral attitude, product attributes, subjective norms and control of perceived behavior affect the intention to purchase ethical apparel to varying degrees. Behavioral attitude has the most significant impact on the purchasing intention, followed by subjective norms, product attributes and control of perceived behavior in decreasing order of the significance level of their impact.

8 Conclusions and recommendations

All the hypotheses made for the theoretical model of this study, namely, H1, H2, H3 and H4, all hold water. That's to say, the four latent variables, i.e. behavioral attitude, subjective norms, product attributes and control of perceived behavior, all have significant positive impact on the intention to purchase ethical apparel. Specifically, behavioral attitude has the greatest impact on ethical apparel consumption behavior, followed by subjective norms, product attributes and control of perceived behavior in decreasing order of the significance level of their impact.

From the finding of this study, we can see that the behavioral attitude of consumers has the greatest impact on behavioral intention. Therefore, we must deepen consumers' knowledge and understanding of ethical apparel and communicate and popularize ethical apparel-related knowledge to foster positive behavioral attitude on the part of consumers. In doing so, we must take the feelings of consumers into consideration, so that consumers can obtain more information about ethical apparel and gain deeper understanding of ethical apparel by osmosis. From the study finding, we can also see that product attributes also play a significant role in influencing behavioral intention. Therefore, the foregoing actions can serve to disseminate environmental protection concepts to consumers, too, thus enhancing their awareness of environmental protection, creating an atmosphere in favor of environmental protection in society, and establishing influential reference groups that play exemplary roles. They enhance consumers' behavioral intention by strengthening their behavioral attitude and subjective norms.

In our study, consumers express their willingness to support fashion-related ethical business yet their shortage of related knowledge. Therefore, they could not accurately identify ethical fashion products in shopping. Moreover, consumers who had actually experienced ethical consumption were quite few. Therefore, providing consumers with relevant guidance and education is especially important for improving their awareness of fashion ethics. For this reason, apparel enterprises must make full use of various media to conduct more effective communication with consumers and make correct releases and disclosures of information on social responsibility to targeted consumer groups through certain channels on a regular basis. This will significantly lower the information identification cost on the part of consumers and in turn promote actual ethical purchasing.

In addition, this study finds that consumers are willing to pay premium prices for ethical fashion products and the premium rate acceptable to them ranges from 11% to 15%. Therefore, retailers must ensure that their ethical fashion products are not priced at a level 15% higher than that of similar products. This finding can provide ethical fashion retailers with relatively good guidance in pricing their products.

Finally, this study finds that consumers pay special attention to the impact dealt by textile and apparel enterprises on environment and that the environmental protection awareness of the general public has significantly heightened. People have realized that they must take the path of sustainable development featuring the harmony between human and nature. It can be said that Chinese consumers tend to attach greater importance to the positive role of ethical fashion products in relation to environmental protection. This points out the right direction for retailers in doing ethical marketing: their product release events and advertisement must highlight the effect of environmental protection wherever possible, so as to strike a chord with their consumers.

Acknowledgements

This paper is funded by the project: General Project of Social Science Program of Beijing Municipal Education Commission (No.: SM202010012001, Name: New Retail Driven Clothing Enterprise Business Model Innovation Research).

References

1. X M Deng, Study on Ethical Purchasing Intention of Consumers under Chinese Scenarios-Based upon TPB Perspective, in Nankai Business Review,**15**(3),22-32 (2012)
2. T H Stone, I.M. Jawahar, J.L. Kisamore, Using the Theory of Planned Behavior and Cheating Justifications to Predict Academic Misconduct, in Career Development International, **14**(3),221-241 (2009)
3. M A Elliott, C J Armitage, C J Baughan, Using the Theory of Planned Behavior to Predict Observed Driving Behavior, in British Journal of Social Psychology, **46**(1),69-90 (2007)
4. H M Kang, D R Fortin, Y J Hyun, Y Eom, Effects of Perceived Behavioral Control on the Consumer Usage Intention of E-coupons, in Psychology & Marketing, **23**(10),841-864 (2006)
5. C Joergens, Ethical fashion: myth or future trend in Journal of Fashion Marketing and Management, **10**(3),360-371 (2006)
6. W D Lu, SPSS Statistical Analysis (Publishing House of Electronics Industry, Beijing, 2003)
7. C.Armitage, M.Conner. Efficacy of The Theory of Planned Behavior: A Meta-analytic Review, in British Journal of Social Psychology, **40**(4),471-499 (2001)

The Effect of University Context on Academics' Engagement with Industry

Zhiyan Zhao*, *Zhi* Chen

[1]Library, Northwestern Polytechnical University, Xi'an, Shaanxi Province, P.R. China
[2]School of Management, Northwestern Polytechnical University, Xi'an, Shaanxi Province, P.R. China

Abstract. Drawing on the theory of planned behavior, this study investigates the impact of academics' perception of university context on their engagement with industry via the mediation of academics' intention to industrial collaborations. The third mission and supportive policy were put forward as two university influence factors. From analysis results of 564 questionnaire responses, we found that academics' perception of university context had significant effect on academic engagement. In addition, the relationship between university mission, supportive policy and academics' industrial engagement were partially mediated by academics' intention to industrial collaboration.
Keywords: University context, The third mission, Supportive policy, Academic engagement.

1 Introduction

University knowledge was widely considered as an important source to promote social and economic development. Various interaction types between university and industrial sectors were studied in previous research. Most of them focused on commercialization which refers to transfer of university generated intellectual property. However, academic engagement, containing consulting, joint research, contract research and informal collaborations such as networking and training, has not gotten enough scholarly attention yet, especially in China [1]. Most Chinese universities, like many others in developing countries, are not considered as emerging technology sources by firms in the global scope. Thus, commercial activities in Chinese universities probably does not play same role as western countries. On the contrary, academic engagement may contribute to catch the full picture of academic knowledge diffusion in China.

In order to solve the above problem, the object of this paper is to discover how scientists' perception of their university context will influence their engagement with industry. Based on pervious study constructed under Chinese context [2, 3], we put forward the third mission and supportive policy context as two university-level factors. Furthermore, based on theory of planned behavior, we will study mediating mechanism of scientists' intention between the relationship of university context and academic engagement.

* Corresponding author: zhaozhiyan@nwpu.edu.cn

 https://doi.org/10.1051/shsconf/20219602006

In the following parts, this present paper begins by hypotheses development. Description of sample collection and methodology is followed in Section 3. Regression results of the hypothesized relationships is presented in Section 4. Discussion, limitation and future research directions are presented in last section.

2 Hypotheses development

2.1 University context and academics' industrial engagement

Scientists' perception of university context will affect their industrial activities. University's third mission means university takes knowledge transfer as a key objective and as an important aspect [4]. Universities with high third mission usually encourage scientists engage in more applied research. Thus, scientists who perceive their universities stressing third mission, would more likely generate applied research outcome. The mismatch between university's supply and industry's demand will be reduced [5]. Hence, scientists would be more likely engage in industrial activities.

Previous studies have found that university's incentive policy, such as sophisticated TTO (technology transfer office) functions, the distribution of benefits between scientists, university and industry sectors and promotion process, promote scientists' industrial activities [6,7]. Sophisticated TTO (technology transfer office) functions could strengthen scientists' advantage to search external collaboration with industry partners in a competitive environment. Incentives could stimulate scientists to engage in industry collaborations for commercial benefits or career development. Based on the above analysis, we put forward the following hypotheses:

H1a: Scientists' perception of university third mission is positively related to their industrial engagement.

H1b: scientists' perception of university supportive policy is positively related to their industrial engagement.

2.2 University context and scientists' intention to industrial engagement

University context could positively affect scientists' intention to industrial collaborations by enhancing their ability and confidence in university-industry interactions. Scientists in universities stressing the third mission would have advantage in applied value research outcome. University incentives could increase scientists' intention to collaborate with industry for commercial benefits or promotion. Sophisticated TTO service could enhance scientists' ability to finish contracts from industrial partners. Thus, both university third mission and supportive policy context probably improve scientists' intention to industrial activities. We put forward following hypotheses:

H2a: Scientists' perception of university third mission is positively related to their intention to industrial engagement.

H2b: Scientists' perception of university supportive policy is positively related to their intention to industrial engagement.

2.3 The mediating effect of scientists' engagement intention

The theory of planned behavior provides the mechanism of relationship between intention and individual's behavior. As the theory of planned behavior, intention is the internal reason that leads to one's behavior [8]. This is because intention is a signal indicating how much one is willing to try and make efforts for an object. A strong intention will result in high possibility of behavior occurrence. Scientists' intention to industrial activities represent for their willingness to academic engagement. It reflects scientists' acceptance to industrial collaborations. The stronger intention scientists hold, the more likely they would engage in industrial activities. Thus, we put forward the following hypotheses:

H3: Scientists' intention to industrial engagement has a positive impact on academic engagement.

Combining the above analysis, we could conclude that supportive university context would enhance scientists' intention to engage in industrial collaborations. Scientists' intention will result in more practice in industry by willingness and efforts. Thus, we put forward the following hypothesis:

H4: The relationship between university context and academic engagement is mediated by scientists' intention to industrial engagement.

3 Methodology

3.1 Sample

Respondents of this study are scientists from Chinese universities. Most of data was collected through email addresses which were searched from universities' official website. Small part was collected through paper questionnaire and mobile tools. To ensure response rate, data collection was conducted in two stages. First, a pilot investigation in a group of 50 scientists was conducted via email which aimed to consult scientists' suggestion of questionnaire. Second, extensive survey was initiated through scientists' email addresses. A reminder mail was sent to respondents who did not reply in one week. Finally, 564 complete data were collected in the final sample. Result of T-test show that early and late returned data have no significant differences in terms of gender, number of publication, rank and industrial connection, suggesting that non-response bias was not problem in our sample.

3.2 Measures

3.2.1 Dependent variable

We used two methods to measure scientists'' engagement with industry by experience of previous study [9]. The first way was to construct academic engagement index by frequencies (above 10 times; 6–9; 3–5 times; 1–2 times; and 0 times) of scientists in 11 types of industrial activities in 2014 and 2015 (see table 1). Academic engagement index was computed through several processes. Difficulty degree of each type of activity was firstly computed: $d_j = 1 - \frac{\sum_{n=1}^{N} b_{n,j}}{N}$, where $j = 1 \cdots 11$ was type of activity; $N = 564$ was the size of our sample; $b_{n,j}$ equaled to 1 if the nth scientist engaged in j th activity, otherwise 0. Second, academic engagement index was computed: $E_n = \sum_{j=1}^{11} d_j T_j$, where T_j was the averages of each frequency interval which were 0, 1.5, 4, 7.5 and 10 respectively. Another measurement of academic engagement aimed to check the robustness of our main result, and was calculated by the number of types scientists engaged in 11 activities.

Table 1. Scientists' industrial engagement activities and difficulty degree.

How Frequently Have You Been Engaged in the Following Types of Activities in 2014 and 2015?	Difficulty Degree
1. A new joint research agreement sponsored by industry (original research work undertaken by both partners)	0.213
2. A new joint research agreement sponsored by a third party (original research work undertaken by both partners)	0.436
3. Work with industry personnel that resulted in a patent or copyright	0.628
4. A new consultancy agreement (no original research undertaken)	0.298
5. Secondment to industry (short- or long-term)	0.543
6. Training of company employees (through course enrollment or temporary personnel exchanges)	0.498
7. Postgraduate training in the company (e.g., joint supervision of PhDs)	0.417
8. A new contract research agreement (original research work done by the University alone)	0.418
9. Creation of new physical facilities with industry funding (e.g., new laboratory, other building on campus)	0.640
10. Attendance at conferences with industry and university participation	0.324
11. Attendance at industry-sponsored meetings	0.427

3.2.2 Independent, mediating and control variables

Independent variables of this study are the third mission and supportive policy context. The scale of third mission was adapted from Guerrero and Urbano [4] and was measured by five items. The scale of supportive was adapted from Wu [2] and was also measured by five questions. Scientists' intention to academic engagement was measured by asking them how likely they would engage in each kind of industrial activities in future.

In addition, this study controlled for several variables that may have confounding effects on academic engagement. A dummy variable was utilized to measure gender (0=female, 1=male). Master and doctoral supervisor were measured by dummy variables, in which master or doctoral supervisor was coded 1 and otherwise was coded 0. Publication and public funds were measured by the number of respondents' publication and public funds in 2014 and 2015. A dummy variable was used to measure academic rank (1= professor or associate professor; 0 = lecturer or assistant). We used a dummy variable to measure industrial connection, in which having a secondary position in industrial departments was coded 1 and otherwise was coded 0. Scientists' previous experience in industrial engagement was scaled by a Likert five-point scale in which 1-5 denoted from totally disagree to totally agree.

3.3 Descriptive statistics results and empirical evaluation of measurements

Descriptive statistics results and correlation matrix of all variables is presented in Table 2. Reliability and validity of measurements were checked. Results of reliability analysis showed that Cronbach's of the third mission, supportive policy and intention were 0.826, 0.835 and 0.955 respectively, indicating the reliability of measurements. Exploratory factor analysis showed that KMO was 0.887 and the significance of Bartlett's test was 0.000, suggesting that it was suitable for factor analysis. The minimum standardized factor loading was over 0.5. Thus, validity of measurements was confirmed.

Table 2. Descriptive statistics results and correlations matrix (N=564).

Variables	Correlation Matrix												
	(1)	(2)	(3)	(4)	(5)	(6)	(7)	(8)	(9)	(10)	(11)	(12)	(13)
(1) Gender	1.00												
(2) Master supervisor	0.14	1.00											
(3) Doctoral supervisor	0.05	0.45	1.00										
(4) Publication	0.08	0.24	0.37	1.00									
(5) Public funds	0.13	0.30	0.28	0.45	1.00								
(6) Rank	0.22	0.42	0.39	0.24	0.31	1.00							
(7) Industrial connection	0.14	-0.03	0.08	0.14	0.14	0.03	1.00						
(8) Experience	0.03	0.06	0.09	0.14	0.07	0.10	0.22	1.00					
(9) Third mission	0.11	0.17	0.12	0.14	0.20	0.22	0.13	-0.10	1.00				
(10) Supportive policy	0.06	0.15	0.11	0.10	0.14	0.20	0.10	0.01	0.32	1.00			
(11) Intention	0.17	0.25	0.19	0.24	0.26	0.36	0.22	0.20	0.43	0.40	1.00		
(12) Academic engagement index	0.22	0.29	0.28	0.31	0.32	0.36	0.31	0.18	0.43	0.36	0.47	1.00	
(13) Academic engagement II	0.26	0.32	0.28	0.32	0.39	0.53	0.24	0.14	0.51	0.36	0.56	0.81	1.00
Mean	0.62	0.52	0.19	6.34	1.38	0.56	0.24	3.17	11.35	10.48	11.97	8.59	6.16
Standard Deviation	0.49	0.50	0.39	5.52	1.24	0.50	0.43	1.05	2.77	2.97	4.08	8.33	3.88

4 Results

Hierarchical regression analysis was used for hypotheses testing. Results were presented in table 3. Model 1 presents effects of control variables on intention. University level factors were added in model 2 to test the effects of third mission and supportive policy on intention. Coefficients were positive and significant indicating that H2a and H2b were confirmed. Model 3 was to confirm the effect of control variables on academic engagement index. In model 4, university factors were added and results showed that H1a and H1b were supported. Model 5 was set to test the effect of scientists' intention on their industrial engagement behaviors and results showed that H3 was supported. Both control variables, mediating and university level factors were put in model 6. Results showed that intention had a partly mediating effect between relationship of university context and academic engagement. Thus, H4 was supported.

Table 3. Hierarchical regression analysis results (N=564).

Variables	intention		academic engagement index			
	Model 1	Model 2	Model 3	Model 4	Model 5	Model 6
Gender	0.536	0.447	1.794^{**}	1.616^{**}	1.478^{*}	1.493^{**}
	(0.327)	(0.291)	(0.634)	(0.577)	(0.606)	(0.573)
Master supervisor	0.843^{*}	0.486	1.928^{**}	1.291^{*}	1.430^{*}	1.157
	(0.366)	(0.327)	(0.709)	(0.647)	(0.680)	(0.643)
Doctoral supervisor	-0.372	-0.281	1.217	1.386	1.436	1.464
	(0.470)	(0.419)	(0.913)	(0.830)	(0.871)	(0.823)
Publication	0.065^{*}	0.054	0.178^{**}	0.156^{**}	0.140^{*}	0.142^{*}

	(0.033)	(0.029)	(0.063)	(0.057)	(0.060)	(0.057)
Public funds	0.269	0.126	0.727*	0.466	0.568*	0.431
	(0.144)	(0.129)	(0.280)	(0.255)	(0.268)	(0.253)
Rank	2.052***	1.401***	3.152***	1.989**	1.940**	1.603*
	(0.362)	(0.327)	(0.703)	(0.647)	(0.689)	(0.652)
Industrial connection	1.546***	0.947**	4.619***	3.527***	3.706***	3.266***
	(0.379)	(0.341)	(0.735)	(0.675)	(0.711)	(0.674)
Experience	0.456**	0.670***	0.578*	0.991***	0.309	0.806**
	(0.150)	(0.136)	(0.291)	(0.268)	(0.280)	(0.272)
Third mission		0.418***		0.833***		0.717***
		(0.054)		(0.107)		(0.112)
Supportive policy		0.337***		0.513***		0.420***
		(0.049)		(0.097)		(0.100)
Intention					0.590***	0.276**
					(0.079)	(0.083)
Constant	7.518***	-0.423	-0.594	-14.897***	-5.033***	-14.780***
	(0.545)	(0.820)	(1.057)	(1.623)	(1.168)	(1.609)
Adjusted R^2	0.210	0.373	0.287	0.412	0.352	0.422
ΔR^2	0.210	0.163	0.287	0.125	0.065	0.135

Note: * $p < 0.05$,** $p < 0.01$,*** $p < 0.001$.

In order to test robustness of this study, we used second measurement of academic engagement to substitute academic engagement index in model 3-6. Poisson regression run by Stata 13 showed that main results remained. Thus, robustness was confirmed.

5 Discussion, limitation and future research directions

Taking Chinese scientists as sample group, this study drew on the theory of planned behavior and used hierarchical regression analysis to study the effect of organizational factors on academic engagement. Our analysis results found that perception of university third mission and supportive policy have significant effect on scientists' engagement with industry. Scientists' intention to industrial activities has a partly mediation role between the relationship of university context and academic engagement.

Theoretical contribution of this study lies in three aspects. First, our research constructed under Chinese context and university-level factors were put forward by analysis of Chinese context. Thus, we bring a novel insight of university-industry interactions in Chinese universities. Second, we broaden extant research of university-industry interactions in China. Previous research mainly focused on commercialization and few concern non-commercial and informal interactions between university and industry. Our research provides a full picture of Chinese university knowledge diffusion. Third, we revealed the mechanism of the impact of scientists' perception of external context on their industrial behavior.

Our study also has practical contributions. Manager and policy-maker should not only focus on formulating relevant policies, but also should concern how scientists perceive external context. Because it is also important to predict their behavior. In addition, scientists' intention plays a mediation role between university context and academic engagement, so managers could promote scientists to engage in academic knowledge diffusion by enhancing their intention to industrial collaborations.

Limitation of this study also indicates directions for future research. Scientists in some department such as art and language refuse to response our survey for the reason that they did not have industrial activities. Future research could utilize secondary data to analyze academics' industrial activities.

Acknowledgements

This research was financially supported by the Fundamental Research Funds for the Central Universities (3102020JC02) and Development Strategy Research Funds of Northwestern Polytechnical University.

References

1. M. Perkmann,V. Tartari, M. Mckelvey, et al. Research Policy, **42**(2), 423-442, (2013).
2. W. Wu. The Journal of Technology Transfer, **35**(2), 203-224, (2010)
3. W. Wu, Y. Zhou. The Journal of Technology Transfer, **37**(6), 812-827, (2012)
4. M. Guerrero, D. Urbano. The Journal of Technology Transfer, **37**(1), 43-74, (2012)
5. H. Lööf, A. Broström. The Journal of Technology Transfer, **33**(1), 73-90, (2008)
6. R. P. O'shea, T. J. Allen, K. P. Morse, et al. R&D Management, **37**(1), 1–16, (2007)
7. R. P. O'shea, H. Chugh, T. J. Allen. The Journal of Technology Transfer, **33**(6), 653-666, (2008)
8. I. Ajzen. Organizational behavior and human decision processes, **50**(2), 179-211, (1991)
9. B. Bozeman, M. Gaughan. Research Policy, **36**(5), 694-707, (2007)

An Exploration into the Influence of Higher Vocational Students' Psychological Capital on Their Employability

Xiong Li, *Qiaoling* Zhang *, *Yongzhong* Cao, *Jinping* Yang and *Mei* Yang

Honghe Health Vocational College, Mengzi City, Honghe Prefecture, Yunnan Province, China

Abstract. Cultivating higher vocational students' employability is one of the most important goals for higher vocational education. Many researchers address teachers' key roles in cultivation of employability. However, the cultivation of employability relates to students' family environment and psychological capital. Through the collection of demographic data, 6700 higher vocational students were investigated with the Positive Psychological Capital Questionnaire (PPQ) and the College Students' Employability Questionnaire (CSEQ). The results show that psychological capital is significantly correlated with employability ($P < 0.01$); although family environment can not regulate the impact of psychological capital on employability, it works independently as psychological capital does; the self-efficacy, optimism and hope in psychological capital significantly affect the employability of higher vocational students from single-parent families. The improvement of psychological capital is conducive to the cultivation of employability of higher vocational students from single-parent families.

Keywords: Employability, Family environment, Psychological capital, Higher vocational students

1 Introduction

In recent years, with the number of college graduates rising increasingly, the employment pressure of college students is growing rapidly [1]. The employment problem has become the focus of the government, universities and other relevant departments which need to be resolved. Employment is an important issue relevant to people's livelihood. Thus many scholars have discussed this issue from different perspectives and expected to find out the solutions. From the perspective of supply-side reform, some scholars suggested that colleges and universities should strive to improve the employability of college students, so as to help them obtain satisfactory employment results [2]; some scholars proposed from the perspective of government management and enterprise development that the government should take tax measures to reduce the burden of enterprises, so as to provide a good employment environment for college students [3]; some scholars believed that the government should focus on the reform of the supply-demand structure, actively adjust the professional settings of various colleges and universities, and resolve the contradiction

* Corresponding author: 1173321980@qq.com

between supply and demand in the employment market [4]. In sum, from the macro level, the supply-demand structure of employment market determines the quality of the employment situation; from the micro level, the evaluation of employment situation and the trade-off of employment result vary among different people, which depend on the individuals' internal cognition and value orientations toward employment results.

Because of the existence of inclination of benefit maximization and the orientations of social values, individuals can evaluate the employment results under the guidance of social values and the pursuits of their goals, thus the employment result itself has the social attributes and becomes an important indicator to reflect the qualities of individuals' employments. The employment result is a result of combination of multiple factors and levels. It is not only affected by the external employment market, but also relates to the college students' internal abilities, such as employability and so on. Some researches show that the employment result is affected by the internal factors such as individual's employability, professional images, personality traits and professional choices. Among them, employability is one of the most important factors affecting possible employment results. The qualities of individuals' employment results are directly relative with their employability [5]. Employability is the precondition for individuals to obtain employment opportunities and improve their job satisfactions. It is also an important core of workplace competitiveness. The improvement of college students' employability is helpful to solve employment problems, guarantee employment opportunities and obtain satisfactory employment results.

The way to improve the college students' employability has always been the focus of many scholars. The literatures about employability usually concentrated on the structure, measurement and antecedents of employability. For example, X. Gong and his colleagues (2018) divided the employability of college students into four dimensions: professional ability, quasi-professional image, personality compatibility and employment personality [6]; W. Cheng (2016) proposed a six-dimensional model of employability through interviews and factor analysis [7]; Y. Cui (2017) discovered a significant impact of psychological capital on employability by analyzing 289 medical students' employability [8].

According to previous researches, psychological capital is the general name of all the positive mentality states which is superior to human capital and social capital, it can help people obtain competitive advantages and improve their employability [9,10]. Individuals with higher psychological capital can obtain more psychological resources, their academic performances and interpersonal relationships are better than those who own lower psychology capital [11]. These essential factors are conducive to the improvement of employability.

The psychological capital is closely related to the individual's family growth environment, and the family environment has an important impact on the individual psychological capital. When individuals get more spiritual and material supports from their families, their psychological needs are better met leading to the growth of psychological capital, which is conducive to the development of employability. These kinds of individuals usually show strong resilience and stable psychological qualities in their careers; however, when people are in the single-parent or divorced family environment, their mental health levels are usually low [12], and their psychological capital is poor [13], which may negatively affect the development of employability. Theoretically, the single-parent family environment may be one of the most important variables among various demographic variables affecting employability.

In order to elaborate the impacts of family environment and psychological capital on higher vocational students' employability, this paper takes the operational definition of demographic variable of family environment as whether higher vocational students come from single-parent families or not, and conducts research on the group of higher vocational students. This paper proposes hypothesis 1: family environment and psychological capital significantly affect higher vocational students' employability. However, there are two modes of the influence of family environment and psychological capital on employability: one is that family environment and psychological capital independently affect employability, there is no interaction between them; the other is that family environment influences employment ability by regulating psychological capital, and there is interaction between them.

The traditional view is that family environment has a great influence on individual employability. Not only dose family environment affect employability directly, but also affects employability through other variables, such as psychological capital. However, there is another view that the influence of family environment on individual's ability is gradually weakened with the increase of individual's social experience, and the socialization process weakens the influence of original family on individual's ability development. With the increase of individual social experience, the development of employability will gradually get rid of the restriction of family environment, so there is no regulatory effect of family environment on employability. In order to find out if there is an intermediate mechanism between family environment and employability, this paper establishes control model and regulation model to verify these views.

In addition, due to the negative impact of family environment, higher vocational students from single-parent families tend to have inferiority complex, jealousy and other negative emotions, which are easy to hinder the developments of their employability. Therefore, in order to better help vocational students from single-parent families develop their employability, the paper sets up hypothesis 2 (Hypothesis 2: psychological capital and its dimensions have a significant impact on the employability of higher vocational students from single parent families). From the different dimensions of psychological capital, this paper provides useful suggestions about the cultivation of employment ability for higher vocational students from single parent families.

2 Object, tools and measurement process of investigation

2.1 Investigation object

A total of 6752 students from a medical vocational college were surveyed by cluster sampling method. 6658 pieces of data are valid for the further analysis. In the data, there are 1436 boys (21.6%) and 5222 girls (78.4%); there are 3884 Han students (58.3%), 306 Hani (4.6%), 1280 Yi (19.2%), 196 Hui (2.9%) and 992 other ethnic minorities (14.9%); there are 5681 vocational students (85.3%) from rural areas and 977 students from cities and towns (14.7%); there are 1015 students from singleton families (15.2%) and 5643 students from multiplex families (84.8%). there are 736 students from single-parent families (11.1%) and 5922 students from non-single-parent families (88.9%).

2.2 Positive psychological capital questionnaire (PPQ)

The Positive Psychological Capital Questionnaire was used to measure the psychological capital of higher vocational students. The questionnaire is a measurement tool developed by Zhang Kuo and his colleagues on the basis of Psychological Capital Questionnaire (PCQ) compiled by Luthans et al. (2010) [14]. The questionnaire has 26 items and 4 dimensions, which are self-efficacy, resilience, hope and optimism. Each item was evaluated by Likert scale. The subjects were asked to choose from "completely inconsistent" to "fully consistent" according to the actual situation. Each item was scored from 1 to 7. The overall Cronbach's alpha coefficient of the questionnaire is 0.90, and the Cronbach's alpha coefficient of each dimension is 0.75, 0.75, 0.90 and 0.80 respectively.

2.3 College students' employability questionnaire (CSEQ)

The employment ability of higher vocational students was measured by the College Students' Employability Questionnaire (CSEQ) compiled by W. Cheng. The questionnaire consists of 89 items, each item is scored by Likert scale ranged from 1 to 7. The scale is divided into six dimensions, which are general vocational skills, professional core quality, career development potential, professional personality, team quality and leadership quality [15]. Previous studies have shown that the questionnaire has good reliability and validity, the internal consistence reliability of each dimension is between 0.83 and 0.97, and the convergence validity is greater than 0.8 [16]. Based on the measured data, the Cronbach's alpha coefficient of the scale is 0.97.

2.4 Measurement process

The unified instruction was carried out in class. The professional staffs were required to carry out measurement in batches, at different times and in different places, ensured the anonymity of the subjects and strictly controlled the common method deviation included in the measurement process [17]. SPSS18.0 software was used to analyze the measured data. The research methods included Pearson correlation, Moderated Multiple Regression and Stepwise Regression.

3 Results

3.1 The relationship between psychological capital and employability

The data of 6658 vocational students were analyzed by Pearson correlation analysis. The results show that there is a significant correlation between the total score of psychological capital and its dimensions, the correlation coefficient is between 0.495 and 0.866; the total score of psychological capital and its dimensions are significantly correlated with employability, the correlation coefficient is between 0.429 and 0.712, showing a positive correlation with high intensity.

Table 1. Pearson correlation matrix of psychological capital and employability.

	1	2	3	4	5
1.Self-efficacy	1				
2.Resilience	0.683**	1			
3.Hope	0.586**	0.503**	1		
4.Optimism	0.561**	0.495**	0.756**	1	
5.Psychological capital	0.866**	0.807**	0.840**	0.827**	1
6.Employability	0.571**	0.429**	0.712**	0.623**	0.695**

3.2 The impacts of family environment and psychological capital on employability

It's possible that family environment and psychological capital affect employability in different ways, that is, family environment independently affects employability or affects it by adjusting psychological capital. When family environment and psychological capital affect employability independently, the control model is established; when family environment affects employability by adjusting psychological capital, the moderated model is established. In order to explore whether the regulatory effect of family environment exists or not, the Moderated Multiple Regression (MMR) equation is established by using the interaction term of family environment and psychological capital [18,19]. Please see the following figure 1.

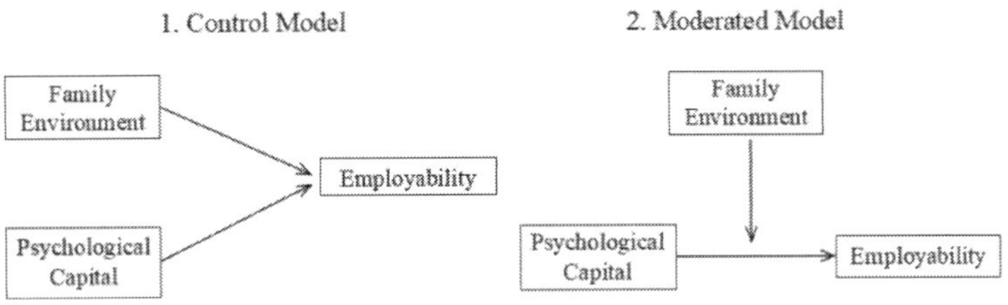

Fig. 1. The Comparison of Control Model and Moderated Model.

As shown in table 2, the explanatory power of psychological capital to employability is 0.735, while the explanatory power of family environment to employability is 0.018, both reaching significant levels, which support hypothesis 1. However, the interaction between psychological capital and family environment upon employability was not significant, $\triangle F(1, 16654) = 0.050$, $p > 0.5$. It shows that family environment can not regulate the effect of psychological capital upon employability, thus the moderated model is not established; but family environment and psychological capital affect employability independently, the control model is established. The control regression equation was $y = 3.431x + 4.532z + 56.553$. Among them, X is psychological capital, Z is family environment.

Table 2. Regression analysis on the influence of family environment and psychological capital on employability.

Model	Independent Variables	R^2	$\triangle F$	β	Beta	t
1	Psychological capital	0.540	7828.503**	3.432	0.735	88.479**
2	Psychological capital	0.541	4.448*	3.431	0.735	88.493**
	family environment			4.532	0.018	2.109*
3	Psychological capital	0.541	0.050	3.486	0.747	14.052**
	family environment			7.753	0.030	0.532
	Interaction term			-0.029	-0.017	-0.225

3.3 The influence of psychological capital on the employability of higher vocational students from single-parent families

Y.J. Wang believes that psychological capital has a significant impact on college students' employability [20]. However, this study does not focus on subjects' family environment. As mention before, family environment is the key variable leading to people's various employability, thus family environment must be considered in further analysis. We select 736 higher vocational students from single-parent families as the research subjects, and test the influence of psychological capital on the employability of vocational college students from single-parent families by Stepwise Regression method.

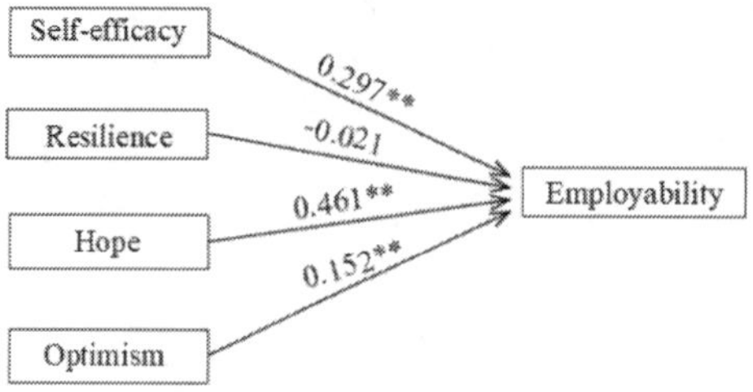

Fig. 2. The Effect of Dimensions of Psychological Capital on Employability.

The results show that psychological capital can still significantly affect the employability of vocational college students from single-parent families. After controlling the effect of each dimension of psychological capital, self-efficacy, hope and optimism can significantly affect employability, their standardized regression coefficients are 0.297, 0.461 and 0.152 respectively, reaching a significant level; however, resilience dimension cannot affect employability, which is different from previous research conclusions [20], which may be caused by different research subjects. Previous studies mainly focus on ordinary college students, but for higher vocational students from single-parent families, their resilience levels are likely to be different from the overall levels of college students. On the whole, the four dimensions of psychological capital explain 60.10% variation of employability, which show that the influence of psychological capital on the employability of vocational college students from single-parent families is significant.

4 Conclusions

4.1 The control model of family environment and psychological capital affecting vocational college students' employability

The research shows that the employability of higher vocational students from two-parent families is significantly better than that of higher vocational students from single-parent families, $F(1, 16655) = 4.448$, $p < 0.05$. As for higher vocational students, a good family environment is conducive to the improvement of employability, while the broken family relationship and the lack of family care are harmful for the development of employability. However, the influence of family environment on employability is limited, and it is not the only factor which determines the development of employability. Compared with psychological capital, family environment has less influence on employability; at the same time, the regulatory effect of family environment on higher vocational students' employability is not significant, which means that family environment can not enlarge or reduce the effect of psychological capital upon employability.

This is inconsistent with the traditional view. The traditional view is that family environment has an important impact on the development of individual ability. As a result of the lack of family care and parental education, the individual growing up in a single-parent family is sensitive, suspicious, and has poor psychological qualities, which seriously restricts the development of comprehensive ability. However, the research shows that family environment and psychological capital are two separated variables that affect employability independently and there is no interaction between them, that is, the influence of psychological capital on individual employability cannot be enhanced by family environment. Family environment is a relatively stable external variable, and its impact on employability will gradually decline with the passage of time. Individuals can heal themselves in the process of socialization, use the cares and helps of peers to make up for the lack of family cares, and repair the negative impacts of the original families on themselves. Therefore, with the increase of people's social experience, their employability will eventually get rid of the constraints of family environment and continue to develop. The impact of family environment on individual employability is generally fixed and short-term, while the impact of psychological capital on individual employability is dynamic and long-term. Psychological capital is a controllable internal variable, which corresponds to the individual's positive attitude, optimistic and progressive personality. Therefore, compared with the family environment, psychological capital has greater impact on the development of individual employability.

4.2 Improving the psychological capital of higher vocational students from single-parent family is conducive to the cultivation of employability

Psychological capital plays an important role in the cultivation of employability of higher vocational students from single-parent families, its pure effect is 0.739. The stronger the individual's psychological capital is, the more active he is to engage in work and study, and constantly improve his comprehensive ability. On the contrary, individuals with weak psychological capital tend to treat work and study with negative attitudes, which are easy to fall into self denial and form a vicious circle [21]. Therefore, giving full play to the positive role of psychological capital is of great significance to personal study, employment and work.

The improvement of psychological capital is beneficial to the cultivation of higher vocational students' employability. Therefore, higher vocational colleges can not only cultivate the students' employability from the perspective of core skills and professional knowledge, but also comprehensively improve the employability of higher vocational students from the perspective of psychological capital. Specifically, it is necessary to carry out the special training of employability from three aspects, such as self-efficacy, hope and optimism. Self-efficacy refers to the individual's belief and effort to complete the challenging tasks. Individuals with high self-efficacy often exhibit full confidence against the tasks and difficulties, and these mental qualities make them concentrate on the tasks without interruptions; hope is that individuals can persevere and adjust to achieve goals; optimism refers to an individual's positive attribution to the present and future [22]. Vocational colleges can focus on the three dimensions, and integrate the cultivation of self-efficacy, willpower training and attribution training into psychological classroom and group activities. By the means of discovering and emphasizing students' own advantages, students' self-efficacy would be cultivated. They can learn to build up their confidences in spite of setbacks by not only accepting their weaknesses, but also learning to see their advantages in the difficult circumstances. This process would benefit their self-confidences and improve their adverse quotients; by the means of helping students set up the reasonable goals that can be achieved through hard work, students would experience the success of self-fulfillment and become more persistent in the challenges. This process would be helpful for students to enlarge their willpower resources; by the means of guiding students to form the favorable habits, students would acquire knowledge effectively and get rid of the constraints of willpower resources which minimize the effect of willpower; by the means of practicing attribution training in psychological class or in extracurricular activities, students would develop their optimisms toward studying and living while they attribute success to competence and attribute failure to effort, their hopefulness about future would be reinforced. In sum, there are many ways to enhance people's psychological capital, through the acquisition of reasonable beliefs, the form of good habits and attribution training in class or in extracurricular activities, higher vocational students' employability would be better fostered.

5 Acknowledgement

This research was financially supported by the 13th Five-Year Plan project of Education Science in Honghe Prefecture, Project No.: 1352127. The title of the project is "Empirical Research on the Influence of Psychological Capital of Students from Single-parent Families on Interpersonal Relationship and Employability."

References

1. P. Wang, F. Ji, China higher medical education, (3),57-58 (2017)
2. Z.C. Wang. Journal of Shijiazhuang Railway University (Social Science), **11**(2), 102-105(2017)
3. L.Y. Zhang, H. Li, X.N. Fu, Modern Business Trade Industry, (29),139-140(2016)
4. R.Y. Liu, L. Wang, T. Chen, Q. Yin, Journal of Guangdong Communications Polytechnic, **18**(2), 118-122(2019)
5. Z.H. Qiao, S. Wang, B.Q. Xie, Z. Wang, Psychological Development and Education, **27**(3), 274-281(2011)
6. X. Gong, T.S. Cai, Jiangsu Higher Education, (1), 91-94, (2018)
7. W. Cheng, Heilongjiang Higher Education Research, (6), 13-15(2016)
8. Y. Cui, Journal of Shangqiu Vocational and Technical College, **16**(5), 85-88 (2017)
9. R.X. Xiao, L.H. Xu., Qiuzhi Guide, (30), 42-43(2017)
10. G.Y. Lin., W.F. Qiu, L.Z. Wei, B.L. Yang, Journal of Jinzhou Medical University(Social Science), **15**(3), 61-64(2017)
11. F.L. Lu, Journal of Kaifeng Institute of education, **37**(7), 161-163(2017)
12. Y. Ge, D. Yan, Journal of Chongqing Jiaotong University (Social Science), **11**(2), 114-117(2011)
13. Q. Tang., M.M. Ye, Journal of Anhui Vocational College of Commerce and Trade (Social Science), **16**(2),62-64(2017)
14. K. Zhang., S. Zhang., Y.H. Dong, Psychology and Behavior Research, **8**(1), 58-64(2010)
15. W. Cheng, Exploration of Higher Education, (5),78-89(2016)
16. W. Cheng, Heilongjiang Higher Education Research, (6), 13-15(2016)
17. H. Zhou, L.R. Long, Progress in Psychological Science, **12**(6), 942-943(2004)
18. H.Z. Qiu, Quantitative research and statistical analysis (Chongqing University Press, Chongqing, 2013)
19. Z.L. Wen, J.T. Hou., L. Zhang, Acta psychologica Sinica, **37**(2), 268-274(2005)
20. Y.J. Wang, Chinese Journal of Applied Psychology, **19**(1),65-71(2013)
21. Z.H. He, Journal of Jiujiang Vocational and Technical College, (4), 55-57(2017)
22. X.Q. Lin, Journal of Wuyi University, **36**(10), 82-87(2017)

Research on the Relationship between the Female Directors' Personal Characteristics and the Quality of Accounting Information

Lu Yin*

School of Economics and Management, Nanchang Hangkong University, Jiangxi, China

Abstract. This paper explores the relationship between female directors and the quality of accounting information. it takes 51 enterprises from 2008 to 2018 as samples, selects the proportion of female directors, age, education background and financial background to study the personal characteristics. The results show that: the percentage of female directors is directly proportional to the accounting information; the higher the female director's education, the higher the quality of information; the female directors with financial background have a specific positive effect on the improvement of accounting information. Based on this, this paper proposes some suggestions on how to strengthen the construction of the board of directors and improve the quality of information about listed companies.

Keywords: Listed companies, Female director, Quality of accounting information, Personal characteristics

1 Introduction

In recent years, the distortion of accounting information of listed companies emerges in endlessly. The root cause is that the accounting information quality is reduced because of the defects of corporate governance structure. As an internal governance organization, this is in a crucial core position of corporate governance. With more and more women appearing in the top management groups, people are increasingly concerned about the effectiveness of female leadership. There are obvious differences between female and male in their solutions and styles. In this context, this paper takes the female directors as sample, discusses the impact of personal characteristics of female directors on the accounting information quality.

2 Literature review

Carter et al. (2003) conducted an empirical study found that the more women on the board of directors, the higher the company value [1]. Ren hao (2010) explored corporate performance is better when there are women in the executive group, and it will be strengthened with the improvement of human capital and social capital of female senior managers [2]. Srinidhi et al. (2011) believe that compared with male executives, female

* Corresponding author: 48261846@qq.com

https://doi.org/10.1051/shsconf/20219602008

executives can bring a management culture that is good at communication, mutual trust and win-win cooperation, so as to effectively reduce the opacity of company information [3]. Zhang Bin (2015) analyze the relationship between the information disclosure about listed companies and. the characteristics of female directors. The research results show that the higher the percentage of female directors, the higher the quality of accounting information. At the same time, the study of female directors has a positive impact on the quality of accounting information [4].

According to the above literature review, the participation of female directors can also urge the fulfillment of corporate social responsibility. At the same time, the quality of statements also reflects the company's sense of social responsibility. The participation of female can improve the accounting information quality, strengthen accounting conservatism, and optimize the level of accounting information disclosure. This paper draws on the research experience of past scholars and judges the quality of enterprise accounting information according to the enterprise information quality rating issued by Shenzhen Stock exchange every year.

3 Theoretical analysis and research hypothesis

The participation of female directors will have an influence on the accounting information quality. This paper intends to analyze the proportion, age, education background and the number of people with financial background in the board of directors, and put forward some assumptions.

(1) The influence of the proportion of female members on the quality of accounting information

Zhao Fang fang and Quan Ya wen (2017) shows that due to the empathy of women and the difference of gender roles, enterprises with more female executives are more inclined to disclose enterprise environmental information [5]. Compared with male directors, female directors pay more attention to details when dealing with the company's affairs, are more cautious in specific business decisions, and are more serious in information collection and processing, all help to improve the operation level of the board of directors. Female directors are more inclined to truthfully reflect the company's financial situation and disclose reliable information. Women have stronger empathy motivation and better cognition of social responsibility. They will pay more attention to the needs of stakeholders. This paper proposed the hypothesis

H1: The number and proportion of female have a positive impact on the accounting information quality

(2) The influence of the age of female on the accounting information quality

The risk preference and business decision-making of board members will change with age and experience. Some studies have shown that young people are more willing to experiment with new opportunities and adapt to the new environment [6]. According to Maslow's theory, the older the board members are, the more socially responsible they are. On the basis of satisfying their basic needs, they are willing to fulfill their social responsibilities to drive their higher needs to be satisfied [7]. In a word, this paper puts forward the hypothesis

H2: the age of female board members has a positive impact on the quality of accounting information

(3) The influence of educational background on the quality of accounting information

According to the theory of human capital, human resource is capital, and its role is closely related to the intensity of investment. The more investment, it will bring greater returns to the enterprise in a certain period of time [8]. Education is the biggest investment. The higher the educational background of the board members, the higher the salary they deserve, the greater the investment of the enterprise, and the greater the final return. Highly educated managers can think about the situation from multiple directions, and make both personal and public interests in the decision-making process. They not only pay attention to the short-term interests of the organization, but also pay attention to the growth of long-term interests of the company, and do not violate laws and regulations and relevant policies. Therefore, this paper puts forward the following assumptions

H3: the educational background of female board members has a positive impact on the quality of accounting information

(4) The influence of financial background of female board members on the quality of accounting information

Board members have different cognitive abilities and different understanding levels of problems, which can feed back to the decisions of enterprises and accounting behaviors. In the process of women performing their duties, whether they have sufficient financial knowledge plays an important role in accounting quality [9]. At the same time, if the audit members in the board of directors have financial professional background, they will have greater supervision on the quality of accounting information, and have a greater right to speak, and their proposals in the decision-making process will have more reference value, so they will be valued and recognized by other board members. To sum up, this paper puts forward the hypothesis

H4: the financial background of female board members has a positive impact on the quality of accounting information

4 Research design

4.1 Sample selection and data sources

This paper randomly selects the data of 51 companies listed from 2008 to 2018 as the samples, according to the following principles: (1) excluding ST and * ST companies, because the financial operation status of these companies is abnormal and the comparability is low; (2) companies with missing or obviously abnormal data exist. After screening, this paper collected 553 groups of samples, excluding 69 groups of samples whose return on assets was blank, and 41 groups of samples whose number of female directors was 0 in the analysis of personal characteristics. Based on this, this paper collected 484 groups of effective samples for empirical study. The selection of dependent variables and control variables in this paper is mainly from CSMAR.

4.2 Variable design

4.2.1 Explained variable

The explanatory variable of this paper is the quality of information disclosure of listed companies on the main board of Shenzhen Stock Exchange. Since 2001, it finally divided the disclosure quality into four grades: excellent (a), good (b), passing (c) and failing (d). this paper uses 3, 2, 1, 0 to assign A, B, C and D respectively according to the evaluation results of information disclosure quality.

4.2.2 Explanatory variables

This paper studies the impact of female directors' personal characteristics on the accounting information quality. In this paper, the personal characteristics of female directors mainly use the proportion, age, education background and financial background to explore the relationship between the accounting information qualities.

4.2.3 Control variable

This paper quotes four control variables: asset liability ratio (Lev), return on net assets (ROA), board structure (DSIZE) and number of board shares (DSH).

Table 1. Symbol definition.

Properties of variables	Variable name	Symbol	Variable definition
Explained variable	Quality of accounting information disclosure	DISC	A is 3, B is 2, C is 1, D is 0
Explanatory variables	Proportion	PER	Number of female directors in board teams
	Age	AGE	Average age of female directors
	Education	EDU	Average educational background of female directors. Among them, technical secondary school or below is 1, junior college is 2, undergraduate is 3, master is 4, doctor and above is 5
	Financial background	BACK	Number of female directors with financial background
Control variable	Asset liability ratio	LEV	Ratio of total liabilities to total assets at the end of the year
	ROA	ROA	The ratio of the net profit of the previous year to the total assets at the end of the previous year
	Board structure	DSIZE	Proportion of independent directors in the board of directors
	Number of board shares	DSH	Shares that held by the board of directors

4.3 Model building

This paper establishes the following model according to the correlation among dependent variables, independent variables and control variables

$DISC=b_0+b_1PER+b_2AGE+b_3EDU+b_4BACK+b_5LEV+b_6ROA+b_7DSIZE+b_8DSH+e$(1)

5 Figures and tables

5.1 Descriptive statistical analysis

Table 2 shows the situation of the quality ratings and the personal characteristics of female directors of 51 listed companies in the past decade.

Table 2. Descriptive statistical results.

	N	Min	Max	Mean	Standard
Quality rating	484	0	3.0	2.041	0.7240
Proportion of women	484	3.85%	39.33%	27.5034%	0.1211842
Age	484	32.33	64.50	47.0749	4.81164
Education	484	1.000	5.000	3.38686	0.766696
Financial background	484	0	8	2.46	2.183
Asset liability ratio	484	0.0437	1.4110	0.561122	0.2026825
ROA	484	-0.187188	0.201387	0.069957	1.3759870
Proportion of independent directors	484	0.29	1.00	0.3644	0.05516
Number of shares held by the board of directors	484	0	584674581	9465576.94	53347630.370
Effective n	484				

From the random sampling, the largest percentage of female directors is 39.33%, the minimum is 3.85%, and the average percentage of women is more than 25%. This shows that in recent ten years, China's enterprise board of directors has gradually paid attention to gender heterogeneity, but compared with the number of men, the proportion is still relatively low, and the number of different enterprises is different. From the perspective of age and educational background, the average age of female directors is 47.07 years old, which is younger. Among them, the oldest company is 64.5 years old, and the youngest is 32.33 years old. From the educational background of female directors, the average educational background is undergraduate and graduate students, which is higher on the whole. According to the number of female directors with financial background, there are 2.46 female with financial background in each company. In addition, from the perspective of asset liability ratio, the average is 56.11%, but the range is from 4.37% to 141.1%, which shows that the gap of asset liability ratio of listed companies is very large. From the perspective of ROA, the average value is 6.99%, but the range is from - 18.71% to 20.13%, which shows that the difference of ROA is large. From the perspective of the proportion of independent directors, the average number is 36.44%, the range is from 29% to 100%, which shows that the proportion of independent directors of listed companies is not big. From the perspective of the number of shares held by the board of directors, the average number is 9465576.94, the minimum value is 0, but the maximum value is more than 500 million, which shows that the level of the number of shares held by the board of directors of domestic listed companies changes greatly.

5.2 Correlation analysis

Table 3. Multiple regression coefficient results.

		Quality Grade	Female Number	Female Proportion	Age	Education	Finance Background
Quality rating	Relevance	1.000	0.726**	0.638**	0.110*	0.703**	0.635**
Proportion of women	Relevance	0.638**	0.741**	1.000	0.111**	0.441**	0.380**
Age	Relevance	0.110*	0.185	0.111	1.000	0.019	0.025
Education	Relevance	0.703**	0.518**	0.441**	0.019	1.000	0.554**
Financial background	Relevance	0.635**	0.609**	0.380**	0.025	0.554**	1.000

"* * significant correlation at 0.01 level (bilateral); * significant correlation at 0.05 level (bilateral).

From the results in the table 3, we can see that some variables in the table have correlation, but the correlation coefficient is not large, so there is no need to worry about multicollinearity between independent variables. The table shows that the proportion of women in the board of directors and the number of female directors with financial background are significantly positively correlated with the company's information quality rating at the level of 0.01, which is consistent with the hypothesis. At the level of 0.01, there is a significant positive correlation between female directors' education and quality rating, which is consistent with the hypothesis. The preliminary results show that gender differentiation of board of directors; high education and financial background of female directors are conducive to the improvement of accounting information quality. However, the average age of female directors is not significantly correlated with the quality rating at the level of 0.05, which is inconsistent with the hypothesis.

5.3 Multiple regression analysis

Table 4. Multiple regression coefficient.

Model		Coefficient of non standardization		Standardization coefficient	T	Sig.
		B	Standard error	Beta		
1	(Constant)	-0.315	0.198		-1.590	0.113
	Proportion of women	0.012	0.002	0.207	5.658	0.000
	Age	0.005	0.004	0.031	1.224	0.222
	Education	0.349	0.030	0.369	11.685	0.000
	Financial background	0.064	0.011	0.194	5.631	0.000
a. dependent variable	\: Quality rating					

This paper adopts the method of multiple linear regression analysis. The regression coefficient is shown in Table 4. From the data in the table, it can be seen that the proportion of women is significantly positively correlated with the quality rating at the level of 0.01, and the non-standard coefficient is 0.012, which indicates that increasing the percentage of women in the board of directors has a positive effect on the improvement of quality rating. The average age and quality rating of female directors is greater than 0.05, which indicates that there is no correlation between them. The age of female directors has little relationship with the accounting information quality. At the level of 0.01, there is a significant positive correlation between the average educational background of female directors and the quality rating, and the non-standard coefficient is 0.349, which indicates that the higher the educational background of female directors, the stronger the sense of social responsibility, and the higher the quality of information disclosure. The number of female directors with financial background is significantly positively correlated with the quality rating at the level of 0.01, and the non-standard coefficient is 0.064, which indicates that the recruitment of female directors with financial background to participate in major decisions is conducive to improving the quality of information disclosure.

5.4 Robustness test

In order to further verify the reliability of the above regression results, starting from the data, this paper adjusts the classification according to different standards, divides the proportion of female directors, age, education background and the number of people with financial background into three levels, and replaces them with "0", "1" and "2", and carries out correlation analysis and regression analysis. The test results are shown in Table 5 and table 6.

Table 5. Results of correlation analysis.

		Quality Grade	Female Number	Female Proportion	Age	Educat -ion	Finance Background
Quality rating	Relevance	1.000	0.608**	0.556**	0.060*	0.550**	0.639**
Proportion of women	Relevance	0.556**	0.654**	1.000	0.120**	0.334**	0.408**
Age	Relevance	0.060*	0.179	0.120	1.000	0.022	0.063
Education	Relevance	0.550**	0.261**	0.334**	0.022	1.000	0.406**
Financial background	Relevance	0.639**	0.488**	0.408**	0.063	0.406**	1.000

"* * significant correlation at 0.01 level (bilateral); * significant correlation at 0.05 level (bilateral).

Table 6. Multiple regression coefficient results.

Model		Coefficient of non standardization		Standardization coefficient	T	Sig.
		B	Standard error	Beta		
1	(constant)	0.999	0.074		13.553	0.000
	Proportion of women	12.338	3.218	0.142	3.834	0.000
	Age	-0.047	0.040	-0.034	-1.185	0.236
	Education	0.528	0.056	0.298	9.450	0.000
	Financial background	0.410	0.043	0.326	9.436	0.000
a. dependent variable \: Quality rating						

From the above table, we can see that the empirical results are basically consistent with the regression results of the original model in the significance level and the direction of correlation coefficient, which proves that the empirical results are robust. Therefore, the results of the robustness test further show that improving the gender heterogeneity of the board of directors and the education and financial knowledge level of female directors have a significant impact on the improvement of the quality of corporate statements.

6 Conclusions and suggestions

(1) According to statistics, the number of enterprises with female directors increased year by year. However, compared with western countries, the proportion of female members in the board of directors in China is still relatively low. And it shows that most female directors are middle-aged women. Most enterprises employ female members with higher education background as directors, which shows that the society pays more and more attention to the knowledge level of managers. From the number of female directors with financial

background, most of them have certain financial background, so as to increase their professional level.

(2) The employment of female members by the board of directors has a positive effect on improving the level of accounting information. it shows that the percentage of female directors is positively correlated with the accounting information quality. This shows that gender differences in board members can play a positive role in guiding the management decision-making, thus improving the quality of enterprise information disclosure.

(3) Female directors with high education level have positive effect on accounting information quality. Higher education can enhance the managers' sense of social responsibility. When disclosing accounting information, they can consider the problems from a long-term perspective. Therefore enterprises should recruit more female board members with higher education, and women should also strive to improve their education and professional knowledge and skills.

(4) Female directors with financial background can improve the quality of accounting information. If female directors have studied in economics and management or have financial experience, they will have a significant positive impact on the quality of enterprise accounting information. In the process of decision-making, if female directors have relevant financial knowledge, they will pay more attention to the issues related to finance, have more voice in business decision-making of accounting information, and their opinions are more representative and professional, which has a positive impact on the quality of accounting information of listed companies.

References

1. C Souza. The Gender and Ethnic Diversity of US Boards and Board Committees and Firm Financial Performance,Corporate Governance: An International Review,**18** (5),60-67,(2010)
2. H Ren. The impact of female participation in top management team on Firm Performance. Nankai management review, **13**, (5),81-91, (2010)
3. F. Gul, B. Srinidhi. Does Board Gender Diversity Improve the Informative of Stock Prices. Journal of Accounting and Economics,**51**, (7), 314-338 (2011)
4. B Zhang. Characteristics of female directors and accounting information quality. Friends of accounting.**23**, (10),84-86, (2015)
5. F F Zhao. Research on the impact of executives' personal characteristics on corporate environmental information disclosure. Accounting news. **27**, (11),100-104, (2017)
6. Y Q Wu. The impact of executives' personal characteristics on the quality of accounting information disclosure: An Empirical Study Based on Listed Companies in information technology industry. Friends of accounting. **30**, (11),63-67, (2018)
7. G L li. Empirical Study on the relationship between corporate governance and accounting information quality. Accounting research. **12**, (2),28-36, (2003)
8. Y Jia. An Empirical Study on the internal and external governance effect of accounting information quality -- Taking Shenzhen A-share listed companies as an example. Accounting and auditing. **24**, (1),138-141, (2017)
9. H B Pan. Vertical concurrent management, property right and accounting information quality. Accounting research. **14**, (7), 19-26, (2016)

Psychometric Performance of Prosocial Tendencies Measure for Undergraduates in Guangdong Province

Yongmei Hou*, *Zhongyi* Zheng and *Que* Zheng

Department of Psychology, School of Humanity and Administration, Guangdong Medical University, Dongguan 523808, Guangdong Province, China

Abstract. Objective: To analyze the psychometric performance of Prosocial Tendencies Measure (PTM) for undergraduates in Guangdong province. **Methods:** Stratified random sampling method was used to select 1109 undergraduates from 10 colleges in Guangdong province.PTM was used to investigate them. Cronbach's α coefficient and split-half reliability were used to analyze the internal consistency of the questionnaire.Convergent validity, discriminant validity and factor analysis were used to evaluate its structural validity. Celling and floor effect were used to analyze its sensitivity. **Results**: Cronbach's α coefficient of the total questionniar was 0.94 and cronbach's α coefficient of 6 dimensions were 0.82～0.87, which met with the requirements of the group comparison. Spearman - Brown split-half coefficient of the total questionniar was 0.95, and the split half reliability of 6 dimensions were 0.83～0.88. The calibration success rate of convergent validity and discriminant validity of the 6 dimensions were all 100%. One component obtained from 23 items which cumulative variance contribution rate was 82.31%. The total score and scores of 6 dimensions were all positive skew distribution, with floor effect in 6 dimensions. **Conclusions:** The psychometric properties of PTM for undergraduates in Guangdong was valid and reliable.
Keywords: PTM, Prosocial Tendencies, Validity, Reliability, Undergraduates

1 Introduction

Prosocial behavior refers to a kind of behavior that the actor voluntarily brings benefits to the recipient (others or society), including helping others, cooperation, sharing, politeness and so on [1]. This kind of behavior usually needs to pay a certain cost or risk, but it is of great significance to individual development and social harmony. Previous studies have shown that prosocial behavior helps individuals maintain positive emotions [2-3], improve academic [4] and professional performance [5], create good interpersonal relationship [4], enhance self-esteem and self-confidence [6], cultivate psychological qualities such as innovation [7], empathy [8] and sense of responsibility [8], reduce behavior problems such

* Corresponding author: 2184456621@qq.com

as aggression [9], internet addiction [10], and promote social adaptation [11-12]. Due to the great significance of prosocial behavior, it is increasingly concerned by all sectors of society.

Prosocial Tendencies Measure (PTM) [13] was developed by American psychologists Carlo and Randall (2002) [13] and revised by Cong Wenjun into Chinese version [14]. It has concise items, clear structure and cross-cultural stability. Since the compilation, it has been more and more widely used. But what is the psychometric performance of PTM for college students? There is no relevant evidence.

Based on the above analysis, this study intends to explore the psychometric performance of PTM for college students in Guangdong Province.

2 Object and method

2.1 Object

2.1.1 Sample Size Estimation

The minimum sample size is calculated by G * power 3 [15]. As PTM is used to evaluate the indicators related to mental health, and disease is one of the important indicators of mental health, we use the prevalence of mental illness of college students to calculate the sample size. Previous studies have shown that the prevalence of mental illness among college students is 9.0-27.7% [16-18], and the test effect value is medium level [16-18], that is, the D value is 0.50-0.80 [19]. In this study, we set the effect value $d = 0.70$, the statistical test power of $1 - \beta = 0.80$, the type I error probability $\alpha = 0.05$, and the minimum sample size is calculated as 678. The minimum sample size is determined as 825 due to a 20% of possible follow-up loss rate.

2.1.2 Sampling

From March to April, 2019, a total of 1200 questionnaires were distributed to college students from freshmen to fifth year from Guangdong Pharmaceutical University, Guangdong Medical University, Guangdong Academy of Fine Arts, Xinghai Conservatory of Music, Jinan University, Guangdong University of Finance & Economics, Guangdong University of Foreign Studies, Dongguan University of Technology, Guangdong University of Technology and Shenzhen University by stratified random sampling. 1109 valid questionnaires were collected, with an effective rate of 92.4%. Among them, there were 536 males and 573 females; 243 in comprehensive colleges, 304 in science and engineering colleges, 145 in liberal arts colleges, 197 in medical colleges, 95 in finance and economics colleges, 74 in sports colleges and 50 in art colleges; 231 freshmen, 219 sophomores, 225 juniors, 230 seniors, 204 fifth year students; 745 from urban areas and 364 from rural areas.

2.2 Tools

2.2.1 Prosocial Tendencies Measure, PTM

It is compiled by Carlo and Randall (2002) [13], and revised by CONG Wenjun [14] into Chinese version. There are 23 questions, divided into six dimensions: "open prosocial behavior", "anonymous prosocial behavior", "altruistic prosocial behavior", "compliance prosocial behavior", "emotional prosocial behavior" and "emergent prosocial behavior".

The Likert 5-point scoring method is used to score from 1 to 5 points corresponding to "not at all" to "completely yes". The higher the total score, the higher the degree of prosocial tendencies.

2.2.2 Self-compiled Personal Information Questionnaire

It includes 4 items, namely, gender, grade, origin, college.

2.3 Collection and arrangement of data

Before the investigation, the researchers who participated in the survey were trained uniformly, and the investigation process and evaluation standard were unified. The consistency test (kappa = 0.81~ 0.90) meet the test requirements.

The questionnaires with scores of more than 50% of the items missing were eliminated. The missing values of the valid questionnaire were estimated and filled with the average. Two researchers independently input the same data using Epidata3.0 software and conduct a unified logic check to ensure the accuracy of the data.

2.4 Statistical methods

Data were exported from epidata3.0 to SPSS 20.0 for statistical analysis. First, the original score of the total scale and each dimension were calculated. The second step was to get the average score and standard deviation of the total scale and each dimension. In the third step, the floor and ceiling effect were evaluated, and then, cronbach's α coefficient and split half reliability were calculated. Finally, convergent validity, discriminant validity and principal component factor analysis were conducted.

3 Results

3.1 The distribution of PTM scores

The ceiling / floor effect is one of the psychological test effects, which refers to the phenomenon that when a test is too simple / complex, the scores of most subjects are close to or reach the upper / lower limit of scores, which makes the evaluation and prediction performance of the test decrease [15]. Table 1 shows that the total score of PTM and the scores of 6 dimensions all tend to be positively skewed distribution. Floor effects appear in all 6 dimensions, but there is no ceiling effect.

Table 1. Descriptive Analysis of PTM (n=1109).

Dimension	X±s	Min	Max	P25	P50	P75	Floor[n(%)]	Celling[n(%)]
open prosocial behavior	8.9±4.9	4.0	20.0	4.0	9.0	13.0	506(45.6)	3(0.3)
anonymous prosocial behavior	12.1±6.9	5.0	25.0	5.0	14.0	18.0	461(41.6)	7(0.6)
altruistic prosocial behavior	11.6±6.3	5.0	24.0	5.0	14.0	17.0	394(35.5)	18(1.6)
compliance prosocial behavior	4.4±2.4	2.0	10.0	2.0	4.0	6.0	488(44.0)	3(0.3
emotional prosocial behavior	9.4±5.2	4.0	20.0	4.0	11.0	14.0	372(33.5)	11(1.0)
emergent prosocial behavior	7.5±4.3	3.0	15.0	3.0	9.0	12.0	523(47.2)	42(3.8)
total score of PTM	54.1±14.7	25.0	98.0	41.0	65.0	89.0	344(31.0)	8(0.7)

3.2 Reliability analysis of PTM

3.2.1 Split half reliability

The 23 items of PTM were divided into 12 and 13 items respectively, and the correlation coefficients of these two parts was 0.93. According to Sperman Brown formula, the split half reliability of the total scale was 0.95. The correlation coefficients of the two halves of 6 dimensions were 0.79, 0.83, 0.81, 0.78, 0.82 and 0.81, and the split half reliability were 0.83, 0.86, 0.88, 0.84, 0.86 and 0.85, respectively.

3.2.2 Internal consistency reliability

Generally speaking, when Cronbach's α coefficient is greater than 0.7, the internal consistency reliability is better. It can be seen from table 2 that the Cronbach's α coefficient of the total scale is 0.94, and those of the 6 dimensions are 0.82~0.87. There is a moderate pair wise correlation among the dimensions (r = 0.01).

Table 2. Cronbach'a Coefficients and Pair Wise Correlation Coefficients.

Dimension	*Cronbach's α*	1	2	3	4	5	6	7
1.open prosocial behavior	.85							
2.anonymous prosocial behavior	.87	.62**						
3. altruistic prosocial behavior	.84	.78**	.68**					
4.compliance prosocial behavior	.82	.63**	.62**	.65**				
5.emotional prosocial behavior	.85	.69**	.67**	.70**	.64**			
6.emergent prosocial behavior	.86	.66**	.68**	.69**	.65**	.70**		
7.total score of PTM	.94	.70**	.71**	.73**	.68**	.73**	.73**	

** P <0.01

3.3 Validity analysis of PTM

3.3.1 Content validity

The correlation coefficient (R) between each item and its dimension is used to represent the convergent validity. Generally, when $R \geqslant 0.4$, it can be considered that the convergent validity is better. Discriminant validity is expressed by the correlation coefficient between the item and other dimensions. If these correlation coefficients are lower than the correlation coefficient between the item and its dimension, the discriminant validity is better. The results show that the correlation coefficient between each item and its dimension was > 0.4, which was higher than those correlation coefficients between the same item with other dimensions. The calibration success rates of convergent validity and discriminant validity were 100%. See Table 3.

Table 3. Convergent validity and discriminant validity of PTM (n=1109).

		Convergent Vality			Discriminant Vality		
dimension	item number	range of *R*	success	success rate	range of *R*	success	success rate
open prosocial behavior	4	.653～.721	4/4	100	.440～.652	20/20	100
anonymous prosocial behavior	5	.689～.726	5/5	100	.612～.703	25/25	100
altruistic prosocial behavior	5	.600～.795	5/5	100	.403～.671	25/25	100
compliance prosocial behavior	2	.772～.775	2/2	100	.614～.703	10/10	100
emotional prosocial behavior	4	.669～.706	4/4	100	.574～.673	20/20	100
emergent prosocial behavior	3	.709～.720	3/3	100	.596～.686	15/15	100

3.3.2 Construct validity

As KMO value is 0.979, Bartlett's spherical test value is 43406. 96 (df= 253), P<0.001. Therefore, the data is suitable for factor analysis. According to the eigenvalue value greater than 1, one principal component is extracted. The factor load of each item is 0.57～0.71, and the cumulative contribution rate is 82.31%.

4 Discussion

We found that the psychometric performance of PTM was good, which was consistent with the results of previous literature [13,14,20-24], suggesting that PTM was suitable for college students in Guangdong.

First, the test of internal consistency reliability and split half reliability showed that PTM had good reliability. The internal consistency reliability of total scale and each dimension was above 0.82; the split half reliability of total scale and each dimension was above 0.83, which was consistent with the results of previous studies [13-14,20-24]. It suggested the split half and internal consistency reliability were good.

Second, there was a high correlation between each dimension and the total score; factor analysis made on the scores of 23 items showed that one principal component was extracted, with the cumulative contribution rate of 82.31%, suggesting that the structural validity of PTM was good. On the other hand, structure extracted from this principal component analysis was not consistent with the 6 dimensional theoretical conception of the original scale [13], which may be caused by different cultural backgrounds. The prosocial behaviors included 6 situations described in the original scale [13]. In essence, they are restricted by personality (especially values). Therefore, there is a significant positive pair wise correlation between the scores of 6 dimensions of the original scale and only one principal component can be extracted.

Third, the scores in all dimensions were positively skewed. Floor effect appeared in 6 dimensions, but there was no ceiling effect. It was consistent with the results of previous literature [14], suggesting that there were still some college students (33.5-45.6%) who were not accustomed to prosocial behavior. Most of them were close to or reach the lower limit of the scores in 6 dimensions at the same time, so the sensitivity was affected to a certain extent.

References

1. N Eisenberg and P A Miller. *The relation of empathy to prosocial and related behaviors*, Psychological Bulletin, *101*(1), 91-119, (1987)
2. F Sun, L Z Lin, M X Dai, M F Luo, G H Cuo, W H Yang, J Jing. *Autistic traits, empathy and theory of mind among college Students with different level of Prosocial behaviors*, Journal of Sun Yat Sen University (Medical Sciences Edition). **38**(6), 938-943 (2017)
3. Y G Li, D Luo. *Research on the positive emotion experience of cooperative prosocial behavior in college students*, Journal of Chongqing University (Social Science Edition). 19(4), 177-182 (2013)
4. M F Wang, H C Chen. *The relationship between adolescents' academic achievement, prosocial behavior and peer acceptance and rejection*, Journal of Psychological Science. 26(6), 1130-1131 (2003)
5. Z Zheng, M Jia. *Relationships between daily affect and customer-oriented service behavior at work: The moderating role of prosocial motivation*, Forcasting. 32(6), 8-14, (2013)
6. A Zuffianò, G Alessandri, B Paula, L Kanacri, C Pastorelli, M Milioni, R Ceravolo, M G Caprara, G V Caprara. *The relation between prosociality and self-esteem from middle adolescence to young adulthood*, Personality and Individual Differences, **63,** 24-29 (2014)
7. X Y Luan, C Chen. The influence of prosocial motivation on innovation behavior: Taking transposition thinking as intermediary variable, Operation and management, **5**, 20-22 (2020)
8. Q Y Li, L S Wang, Y Tian. A study on the correlation between empathy, social responsibility and prosocial behavior among College Students, Modern communication, **13**, 107-109 (2018).
9. Y H Liang, M H Z Hou. *The influence of prosocial tendencies on Implicit Aggression -- the Enlightenment of social learning theory on prosocial behavior of College Students under different levels of violence*, Theory Research, **21**(11) 59-61 (2014)
10. M Q Mo, Y H Yang, X Z Xu. Influence of pro - social motivation on internet addiction: Chain mediating effect of basic psychology needs, Journal of Shaoyang University(Social Science Edition), **22**(5) 114-120 (2015)
11. L K Zheng, X Liu. The college students' prosocial behavior: based on structural equation model of the psychological crisis' intervention, Statistics & Information Forum, **28**(5), 100-106(2013)
12. P Jin, G M Sun, Q M Chen. *Research on the relationship between prosocial behavior and adaptability of college students*, Economic Herald, **23**, 76-77, (2013)
13. G Carlo, B A. Randall. *The development of a measure of prosocial behaviors for late adolescents*, J Youth Adolesc. **31**(1), 31-44 (2002)
14. W J Cong. Research on the types of prosocial behavior of college students, Ma. D. Thesis, Nanjing Normal University (2008)
15. J Cohen. Statistical power analysis, Curr Direct Psychol Sci. **1**(3), 98-101 (1992)

16. Y Hu. *A study on the limitations of SCL-90 in mental health survey of freshmen*, Journal of Changchun University. **22**(4), 444-447 (2012).
17. Y Teng. *The application of cluster analysis and discriminant analysis in the results of SCL-90 among test college students*, Chinese Journal of Social Medicine. 34(5), 460-462 (2017)
18. L Y Shi, J P Xu, ZHOU Lang, WU Xiujuan, ZHAO Huashuo, GENG Debao, ZHANG Xunbao. Reference value, reliability and validity of SCL-90 in college students, Chin J Sch Health. 34(2), 223-224 (2013)
19. H C Zhang, J P XU Jianping. Modern psychology and educational statistics, Beijing Normal University Publishing Group, Beijing, 2012.
20. C Zhang, Y J Zhang, Q Y Wang, L J Hao. *Research on the relationship among post-traumatic growth, gratitude and prosocial behavior*, Journal of Logistics University of PAP (Medical Sciences), **27**(6), 489-494 (2018)
21. W Yao, L S Wang, Hao Li. The effect of college students' shame on prosocial behavior, psychology: Techniques and Applications, **7**(1) 34-38 (2019)
22. Y F Lei, D X Guan, L Chen. Relationship between parental rearing, empathy and early prosocial behavior of adolescents, Youth and Adolescent Studies, **2**, 59-64 (2020)
23. T C Li. The influence of gratitude on prosocial tendency of college students, Journal of Campus Life & Mental Health, **18**(3), 237-240 (2020)
24. Lu Li, B J Ye, L Y Ni, Q Yang. Family cohesion on prosocial behavior in college Students: Moderated mediating effect, Chinese Journal of Clinical Psychology, **28**(1), 178-180 (2020)

A Study of Industry-university-institute Cooperative Education in Colleges and Universities against the Background of Emerging Engineering Education

Weijie Fan[1], *Yangzu* Fan[2,*], *Ju* Zhang[3], *Jianghong* Mao[4] and *Qiang* Li[5]

[1]NingboTech University, Ningbo 315100, China
[2]Zhejiang Post And Telecommunication College, Shaoxing, China
[3]NingboTech University, Ningbo 315100, China
[4]NingboTech University, Ningbo 315100, China
[5]NingboTech University, Ningbo 315100, China

Abstract. With emerging engineering education becoming a new strategic direction of the higher engineering education reform in China, it is an important issue faced by colleges and universities to comprehensively improve their abilities of training talent, conducting scientific research and serving the society. Promoting industry-university-institute cooperation is a key measure for colleges and universities to keep up with the pace of higher education and socio-economic development. Colleges and universities need to improve the industry-university-institute cooperative talent training mechanism, and establish an industry-university-institute cooperative education system based on public technology service platforms, to promote the combination of technology and production through cooperative education based on their current situation of research, push forward the supply side reform of higher education, and provide human resources, technical support and industrial services for social development against the background of emerging engineering education. While enhancing their levels of scientific research and education, colleges and universities can promote social progress and help enterprises create economic benefits, to achieve win-win cooperation with all relevant parties in the society.
Keywords: Industry-University-Institute cooperative, Emerging Engineering Education, Education system

1 Introduction

In order to cultivate high-quality composite talents with innovation ability, engineering practice ability and international competitiveness that meet the needs of emerging industries and economies, the government has proposed an important measure for the upgrading and economic transformation of the Emerging Engineering Education construction. The system

* Corresponding author: fyz@zptc.cn

of industry-university-institute cooperation is a complex structure system with the participation of enterprises, universities, scientific research institutes and social service institutions [1]. Based on the background of emerging engineering education, and in order to adapt to the development of the times, actively integrate into the driving force of social innovation and industrial transformation and upgrading, further promote the structural reform on the supply side of education, and strengthen the cultivation of students' basic theory, innovation spirit and practical ability, universities need to carry out industry university research cooperation with enterprises. It has become the consensus of the whole society to accelerate the construction and development of emerging engineering education, cultivate talents in urgent need of the new economy, and cultivate talents who will lead the development of technology and industry in the future [2]. At the same time, the construction of emerging engineering education brings opportunities and new challenges to the upgrading of the cooperation of production, learning and research in local universities. The industry-university-institute cooperation education is the only way for the development of high-level universities and first-class universities. Because of the continuous innovation of the practice of industry-university-research cooperation education, universities strive to set up a platform for industry university research cooperation education, scientific research and technical service, which provides strong talent support and scientific and technological services for the local economic development and industrial technological progress [3].

Fig. 1. Industry-university-institute cooperation.

2 Improve the industry-university-institute cooperative talent training mechanism

The guidelines for the construction of emerging engineering education clearly put forward that it is necessary to ask for the construction of industries, update the traditional disciplines, and promote the cross combination of existing engineering disciplines; it is necessary to ask for the content of technological development and reform, and update the knowledge system of engineering talents. In order to meet the needs of social development and solve the problem of disconnection between talent training, economy and science and technology, it is urgent to innovate and reform the talent training mode in universities. The cooperative education mode pays attention to the relationship among production, research and learning, which not only increases the cooperation among the three, but also optimizes the function of talent training in universities.

The defect of talent training mode is a great resistance to the development of production, research and learning[4]. Although the education reform has been highly valued by the society, the current part of colleges and universities still follow the pattern of the traditional teaching mode, which is embodied in the emphasis on knowledge teaching and the neglect of the practice link. The students cultivated are difficult to have the ability to solve complex problems [5,6]. The main purpose of professional learning is to complete classroom assignments or experiments, which leads to students' lack of ability to think and explore.

The domestic industry-university-institute cooperative model started relatively late, and colleges and universities can rely on industry-university-institute cooperative to strengthen students' corresponding practical and innovative capabilities. Although the industry-university-institute cooperative model has gradually risen to the national level, it is still not clear how to develop talents based on the actual situation of universities. Colleges and universities should take social needs and professional characteristics as the guidance to set up talent training objectives, so as to improve the fit between talent training objectives and social needs. In this way, we can take full advantages of collaborative innovation of production, learning and research, and construct the training mode of excellent talents. The university should pay attention to the cooperative training of students' basic abilities and professional skills, so as to promote the development of talents towards the application-oriented direction.

2.1 Course teaching

Colleges and universities should not only attach importance to professional knowledge and basic knowledge, but also consider the necessary skills needed by society and employers. Teaching should be carried out with the ability of scientific and technological innovation as the core, which needs to reflect the epochal character of teaching. The society needs colleges and universities to develop an open, advanced and practical curriculum system of collaborative education. In addition, universities should also improve the advanced nature and comprehensiveness of the courses from the elective courses, effectively expand their horizons and cultivate their independent thinking and work ability. The construction of theoretical curriculum system highlighting the cultivation of innovation ability, while making full use of the technical force of front-line production units, introducing their new technology and new equipment in the classroom, to ensure that the content of the curriculum keeps pace with the times.

2.2 Scientific research training

Teachers should bring advanced research methods and analysis methods into the curriculum system, promote the integration of analysis tools and applications, and ensure that students can choose matching coping styles when facing different problems. Students should be encouraged to actively participate in the project discussion and contact the advanced research platform as much as possible, so as to master the advanced scientific research means and improve their scientific research literacy. Establish a new system of senior with junior, let senior students bring junior students into the scientific research work together, junior students preliminarily master the scientific research analysis methods and test methods in the learning process, senior students improve their scientific research ability and stimulate scientific research innovation in the teaching process.

2.3 Innovative practice

The construction of the practical curriculum system that emphasizes the cultivation of practical ability. The biggest difference between emerging engineering education mode and traditional teaching mode is that it pays more attention to the cultivation of students' practical ability [7-8]. The cooperative education mode of industry-university-institute should be used to guide students to participate in project development and production practice for a long time, so as to ensure that they are familiar with the corresponding production process and application needs. Then relying on the guidance of the tutors inside and outside the school, let students clear the direction of the subject, and improve their innovation practice ability. Universities should strengthen the integration of production and learning, make full use of the practice platform of enterprises and institutions, let students learn the required knowledge in the real entrepreneurship and innovation work, make students feel the "real scene" of the work, effectively improve students' lack of understanding of society and occupation, and minimize the employment adaptation period of students[9].

3 Establish an industry-university-institute cooperative education system based on public technology service platforms

University Science Park is an important medium of university technology transfer. Local colleges and universities need to constantly improve the corresponding school enterprise cooperation mechanism in combination with the characteristics of colleges and enterprises, and promote the construction of industry university research cooperation. For school enterprise cooperation, it is necessary to select a matching cooperation mode according to the enterprise situation. Both sides need to clarify the core areas and cooperation mechanism of cooperation, and regularly organize exchange meetings. Colleges and universities can set up off campus practice bases with the help of school enterprise cooperation mode, so that students can go deep into the actual production links of enterprises, clarify the application needs of enterprises, so as to formulate corresponding research directions, and actively promote the transformation of scientific research results into productivity. Universities should build and utilize the technology service platform of academic research cooperation to continuously improve their social service ability.

The school uses the technical strength of enterprises and large-scale instrument test equipment, according to the needs of the core technology research in the real society, and under the cooperation of the government, schools and enterprises, establishes a public technical service platform with complete functions. In the science and technology park, vigorously carry out technical consultation, technical services, technological development and technology transfer. Which will promote the trade, promotion and diffusion of technological achievements and intellectual property rights, and promote technology transfer and transformation of achievements. Based on the public technology service platform, students can participate in relevant work as much as possible, so that students in school can understand the needs of enterprise work, social science and technology development and enterprise engineering. The each major should also establish a dual tutor guidance mechanism, which can employ excellent enterprise talents as external tutors, and establish a matching training program combining with students' cognition and thinking mode. The internal and external tutors should jointly undertake the teaching task.

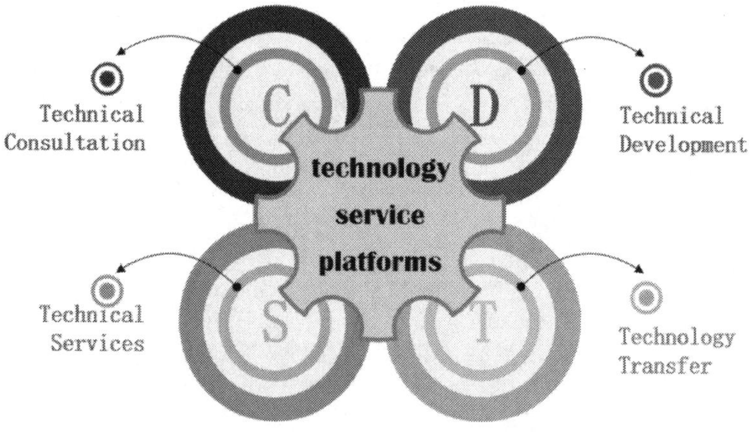

Fig. 2. Functions of technical service platform.

4 Conclusion

The construction of emerging engineering education is still making progress, when colleges and universities should ascertain the status quo of emerging engineering education construction and industry-university-institute cooperation, figure out problems in existing industry-university-institute cooperation among all parties, and present new paths for this cooperation, thereby promoting the reform of the industry-university-institute cooperative education model. Through close cooperation among universities, enterprises and governments, we can improve the teaching system, improve the quality of personnel training, and promote the transformation of scientific research results into productivity. Pay attention to the cultivation of individual ability of students, in order to achieve the purpose of individualized development and comprehensive development of students. Eventually, achieve win-win cooperation with all relevant parties in the society.

Acknowledgement

This research was financially supported by the Zhejiang Province Higher Education "13th Five-year" Teaching Reform Research Project (jg20180442).

References

1. M Zhu, C Chang, L Yu, *Research on construction of harmony evaluation index system of industry-university-research cooperation system,* Journal of Shenyang University of Technology. **12**, 67-72, (2019)
2. S Gao, *Implement the emerging engineering education F-plan and cultivate engineering leaders,* Research in Higher Education of Engineering. **04**, 19-25, (2019)
3. W Liu, M Chen, X Zheng, *Research and Practice of the industry-university-institute cooperative in the New Era,* Chinese University Technology Transfer. **12**, 80-82, (2019)
4. M Gao, Q Wang, C Liu, *Teaching reform and exploration of PLC technology and application course of applied undergraduate colleges under the background of*

emerging engineering education, Journal of Langfang Normal University. **18,** 124-128, (2018)

5. X Chen, W Luo, S Luo, *Research on the upgrading of the emerging engineering education under the background of emerging engineering education*, Education Modernization. **06**, 7-8, (2019)
6. J Tan, L Hua, *Improvement and upgrading of emerging engineering education in local universities under the background of emerging engineering education*, Industrial Innovation. **10**, 238-244, (2019)
7. C Xiu, B Ma, Z Xu, *Reform and Practice of Production-University Integration and Cooperative Education Model for Surveying and Mapping Engineering Specialty under the Background of New Engineering*, Education Teaching Forum. **49**, 85-88 (2019)
8. J Luo, X Li, Y Xi, X Li. *Exploration and practice of innovation and entrepreneurship education reform*, Beijing Education: Higher Education Edition, **10**, 37-39, (2015).
9. Z Wang, Y Shan, S Wang, *Constructing a New System of " Five in One " Innovation and Entrepreneurship Education - - Exploration and Practice of Innovation and Entrepreneurship Education Reform in Inner Mongolia University*, teaching of university of china, **06**, 29-32, (2017)

Early Childhood Education and Care in a Post-pandemic Economy: Lessons from China

Jincheng Ma* and *Lei* Guo

Beihang University, Beijing, China

Abstract. The mismatch between the demand side and the supply side in the Early Childhood Education and Care (ECEC) sector has been a recurring pattern for a long period of time. Unlike the other levels and types of education, over 60% of provision in the ECEC sector comes from private organizations in China. Following the outbreak of the coronavirus pandemic in 2020, challenges have emerged in the high-level customer contact services and the classic labour-intensive industries. From the "regular epidemic prevention and control" point of view, regulatory reform is necessary, such as translating the "price cap regulation" into the "rate of return regulation", enforcing the demand side incentive, combining regulatory innovation and technical innovation, and creating good conditions for the professional development of teachers in the ECEC sector.
Keywords: ECEC, Post-pandemic Economy, Human Capital, Regulation.

1 Introduction

The Early Childhood Education and Care (ECEC) sector is important for the reconciliation of work and family life, especially for developing countries. It will at once be evident that women get more benefit from enlarging the scale of public investment in the early stages, as the gender-equitable labour participation on a full-time and universal basis will reach a higher plane. From the demographic point of view, some researchers argue that the cost of ECEC is invisible and a hard curb on a family's desire to have a baby, since the ECEC sector can influence the direct and/or indirect costs of having children [2]. What's more, the core idea of increasing formation of early stage human capital such as focusing on disadvantaged families, and focusing more on prevention rather than on compensation, has undoubted significance for alleviating poverty in modern society [6].

However, the outbreak of COVID-19 has plunged the global economy into a sudden and unprecedented recession. This crisis will catalyse some huge changes, and few industries will avoid reform, restructure or removal. In this paper, we will review the challenges in the ECEC sector in a post-pandemic economy, taking China as an example, and then put forward several suggestions.

2 The supply side of the ECEC sector in China

* Corresponding author: majincheng0901@gmail.com

Economic transition brought about huge changes in the social service sector in China at the end of the 20th century. In particular, several state welfare systems were abolished in order to raise economic efficiency, however, the provision of social services also declined gradually. In the ECEC sector, approximately 50% of children entered primary school without any pre-school education experience in the first decade of this century in China. This situation was criticized by many scholars in China and abroad, including the 2000 Nobel laureate James J. Heckman [5]. When policymakers in China realized the seriousness of situation, a 10-year-plan was enacted in 2010, causing the enrolment rate to rise sharply. By the end of 2019, the enrolment rate in preschool education reached 83.4%. But, for children aged 3 years and below, enrolment to formal childcare is 4.1%, compared with average 34% in OECD countries [8].

Another characteristic of the ECEC sector in China is that private organizations consist a large proportion of the suppliers. Table 1 below shows that no less than 60% of supplies in the ECEC sector in China are private organizations, which is very different from what situation in primary, secondary and higher education across the country. From the demand side, it may seem that urbanization is underlying reasons for this situation, but in fact the rating of budget priority of the ECEC sector of local governments under the fiscal system of decentralization is key. It is the incentive non-compatibility in the public sector at work.

Table 1. The ECEC enrolment and private ratio in China.

year	Facilities [national] -1	Facilities [private] (2)	Enrolment [national] (3)	Enrolment [private facilities] (4)	private ratio(children) (4)/(3)	p private ratio(facilities) (2)/(1)
2010	150400	102289	29766700	13994700	47.01%	68.01%
2011	166800	115404	34244500	16942100	49.47%	69.19%
2012	181300	124600	36857600	18527400	50.27%	68.73%
2013	198600	133500	38946900	19902500	51.10%	67.22%
2014	209900	139300	40507100	21253800	52.47%	66.36%
2015	223700	146400	42648300	23024400	53.99%	65.44%
2016	239800	154200	44138600	24376600	55.23%	64.30%
2017	255000	160400	46001400	25723400	55.92%	62.90%
2018	266700	165800	46564200	26397800	56.69%	62.16%
2019	281200	173200	47138800	26494400	56.21%	61.59%

Sources: National educational development statistical bulletin. Information on http://www.moe.gov.cn/jyb_sjzl/sjzl_fztjgb.

The third characteristic of the ECEC sector in China is that it is a classic labour-intensive industry. The proliferation of information technology and Artificial Intelligence (AI) in the ECEC sector is much slower than in any other stage of education. As employee costs consist of the major part of the overall cost in the ECEC sector, the ECEC industry is vulnerable to "the Baumol cost disease". What's worse is that the regulation is something like a "Price-Cap", which can enhance productivity in a capital-intensive industry but has the disincentives to invest in a human service industry. For a labour-intensive industry, "inherent in the technological structure of each of these activities are forces working almost unavoidably for progressive and cumulative increases in the real costs incurred in supplying them" [3]. If the regulation authority keeps the price down, while this may succeed temporarily, service quality will diminish in the long run.

3 Some challenges

The coronavirus pandemic has plunged the global economy into a sudden and unprecedented recession. All kinds of occupancies such as cinemas, gymnasiums, and schools have closed following national lockdowns. Businesses who aren't able to make use of solutions such as the potential of cloud computing have lost huge portions of their revenue. In the summer, with the coronavirus gradually under control in China, high-level customer contact services have started to resume business.

Pre-school education has been the last service to resume among all levels and types of education in China. Across the country, most local kindergartens reopened after "the winter vacation" in June. During this "extra-long winter vacation", public organizations have been relatively less affected because their costs are covered by governmental expenditure. However, private organizations are not only unable to collect tutoring fees in advance under policy constraints, but they also need to continue to pay teachers' salaries and housing rents during the winter vacation, thus these organizations have experienced an unprecedented "hardest winter and spring" as the media says.

As many teachers in private organizations have only received partial wages or even no wages at all, a large number of teachers in private organizations have resigned or changed jobs during this period, especially the male teachers. In addition, even after the resumption of work in June, due to the smaller number of children returning to kindergartens, a considerable number of teachers are idle. Teachers who have been laid off or resigned have begun to start their own "micro businesses".

Of course, some teachers in private organizations do not want to end their careers and they are actively seeking new ways of survival. The typical approach of private organizations has been to start "side-line production" whereby some private organizations are using their own catering facilities to organize teachers to start the production of take-out meals; some private kindergartens with ample outside space are even selling barbecue food in the evening and night.

In April, when the resumption of work was still unclear, a survey in western China revealed reduction in student resources, loss of teachers, and rent pressure were the three main impacts of the pandemic on private kindergartens, which had severely hit supply organizations. The loss of teachers brings unfavourable expectations to upstream students in the normal colleges and universities, leading to weak follow-up development.

4 Conclusion and suggestions

From the education point of view, there seems to have been no basic law to abide by in the ECEC sector in China until now, and the mismatch in the demand and supply side in this sector has been a recurring pattern for a long period of time. This mismatch goes even badly after of the coronavirus epidemic outbroke this year. It deteriorated not only many private kindergartens' on the brink of financial catastrophe, but has also had an adverse impact on the industrial chain, for example, some children's products for intellectual development and physical fitness are under-operated. From the macro-economic point of view, the closure of kindergartens has also hindered women from going out for employment, and has adversely affected Chinese women's willingness to bear children. Therefore, legislation on the pre-school stage of education in China is an urgent need.

4.1 Other suggestions

4.1.1 Entry regulation reform

Service organizations in the ECEC sector in China are relatively large and monotonous, and such large kindergartens may not satisfy the heterogeneity of the demand side in China. In contrast, in the United Kingdom, in addition to pre-kindergartens and kindergartens, there are other facilities such as Family Child Care Homes and Group Homes and Group Home Providers [7]. These small-and-micro-providers are scarce in China and applicants to provide ECEC services in the country must meet certain standards such as scale and number of classrooms to be accepted. In fact, mass employment is something concern with the small-and-micro-provider when the policy focus on "ensuring stability in employment, financial operations, foreign trade, foreign investment, domestic investment, and expectations". In the upcoming long-term period of "regular epidemic prevention and control", regulations must be reformed to foster pro-heterogeneity organizations, which can respond with agility to unexpected risks.

4.1.2 Translate the price cap regulation into the rate of return regulation (RORR)

The ECEC service is a kind of credence goods, and is vulnerable to "the Baumol cost disease". Generally speaking, credence goods have the characteristic that though consumers can observe the utility they derive from the good ex post, they cannot judge whether the type or quality of the good they have received is the ex ante one they needed. Moreover, consumers may even ex post be unable to observe which type or quality they actually received. An expert seller, however, is able to identify the type or quality that fits a consumer's needs by performing a diagnosis [4]. When the price of these goods is tightly controlled by the regulatory authority, the quality of the service may deteriorate. In such a context, low-quality organizations can drive out high-quality organizations in the market as they can secretly lower their service cost, thus Gresham's Law also works in the ECEC sector. When turn for the RORR, under the disclosure mechanism, consumers will get what they pay for; and the "Averch-Johnson effect" [1] under the RORR will also attract investment from private sector.

4.1.3 Enforce demand side incentive, especially the taxation incentive

A scientific incentive mechanism will yield twice the result with half the effort. The role of clients in monitoring providers can be strengthened by increasing choice on the demand side and increasing participation in service delivery. Indeed, "when clients are given a choice among service providers, they reveal their demand by 'voting with their feet' " [9]. Some local governments in China subsidize preschool services in the form of vouchers, but offer taxation deduction only for the cost of at age 3-5, and there is no tax credit until now. As the childcare tax deduction and credit are primarily designed to provide local governments financial relief and to avoid a fragmented child care subsidy system, the public funding of the ECEC sector should mainly via the tax method. However, any policy must be credible, as long-term incentives will yield stable expectation.

4.1.4 Technical innovation

In a recent article, the president of a global technology firm asked an important question: "Who will be the winners in a post-pandemic economy?". His analysis was that the success of various types of services hinged on their ability to cope with three key factors: (1) "Businesses that use cloud computing will not buckle under the pressure of the coronavirus pandemic"; (2) "Further automation and artificial intelligence will enhance the resilience of supply chains"; and (3), "Successful businesses will have a combination of resilience and agility." Clearly, the prohibition of online courses impairs the providers' operational capacity in "regular epidemic prevention and control" circumstances. COVID-19 will force a rebirth of many high-level customer contact service industries, as people are forced to remain at home during a lockdown. Thus, regulations must innovate and use the right technology. In the future, ECEC services must rely on the internet and use more artificial intelligence. Perhaps a revolution in the ECEC sector is looming, when childcare services become a distributed system, and providers can service clients remotely.

4.1.5 Finally, authorities in China should create good conditions for the professional development of the teachers.

"Teachers are the foundation of teaching and the source of developing education", and teachers are also core assets and the lifeblood of a kindergarten. However, in recent years, private organizations have faced the dual dilemma of teacher recruitment difficulties and teacher turnover. Generally, heavy workloads, lower salaries and obstacles in career development have been the main causes for teachers to leave the profession. Thus, measures should be focus on several key factors, such as making the starting salaries for teachers among the most competitive in the graduate labour market, building on above-inflation average pay increases for teachers, and ensuring that teachers are in no doubt that the government fully supports them in every stage of their career.

References

1. H. Averch, L. L. Johnson, Am. Econ. Rev., **52**, 1053(1962)
2. G. S. Becker, Conference of the Universities-NBER, Princeton, N.J.: Princeton University Press, 209(1960)
3. W. J. Baumol, W. G. Bowen, Am. Econ. Rev., **55**, 495 (1965)
4. U. Dulleck, R. Kerschbamer, M. Sutter, Am. Econ. Rev., **101**, 526(2011)
5. J. J. Heckman, China Econ. Rev., **16**, 50 (2005)
6. L. Guo, J. Qu, Econ. Perspectives, **116**(2019).
7. E. C. Melhuish, International J. of Child Care and Educ. Policy, **10**, 1 （2016）
8. K. Müller, K. Wrohlich, Lab. Econ., **62**, 1(2020)
9. World Bank, World Bank Report, No. 53746-CN (2011)

Exploration of Innovation and Entrepreneurship Education Model in Higher Vocational Colleges based on Rural Revitalization Strategy

Guohui Su*

[1]Guangdong Mechanical and Electrical Polytechnic, Guangzhou 510515, China;
[2]University Of Chinese Academy of Social Sciences, Beijing 102488, China;

Abstract. The outbreak of the COVID-19 epidemic has had a great impact on economic and social development, the task of poverty alleviation has become more difficult, and the strategy of rural revitalization is facing greater challenges. Higher vocational colleges undertake the important mission of training and conveying advanced skilled talents needed for the country's economic and social development, and are bound to assume their own responsibilities in the strategy of rural revitalization. The innovation and entrepreneurship education for college students promoted by higher vocational colleges in recent years provides a feasible implementation path for higher vocational colleges to participate in the country's rural revitalization strategy. Based on the background of the national rural revitalization strategy, this paper points out the great significance of the integration of innovation and entrepreneurship education in higher vocational colleges into the rural revitalization strategy, and puts forward specific measures to realize the integration of innovation and entrepreneurship education into the rural revitalization strategy.
Keywords: Rural Revitalization, Innovation and Entrepreneurship, Integration, Model

1 Introduction

The proposal of the rural revitalization strategy is a major strategic arrangement and deployment made by the party and the state on the basis of focusing on the overall situation of the cause of socialism with Chinese characteristics and conforming to the people's yearning for a better life in the countryside. It is the grand goal of our country to win a decisive victory and build a moderately prosperous society in an all-round way, as well as a major historical task of building a modern socialist power. It is also an important measure to do a good job in the work of agriculture, rural areas and farmers in the new era.

Higher vocational colleges undertake the task of training and conveying advanced skilled talents needed for the country's economic and social development, and must fulfill the educational mission of higher vocational colleges in the national strategy of agricultural modernization and rural revitalization.

* Corresponding author: 506602681@qq.com

In recent years, higher vocational colleges carry out and promote innovative and entrepreneurial education for college students, attach importance to the reform and innovation of traditional educational methods and methods, and pay attention to the cultivation of students' innovative quality and entrepreneurial skills. especially with the extensive development of Internet + Innovation and Entrepreneurship Competition for Chinese college students, the "Youth Red Dream Tour" has been implemented.

It combines the innovation and entrepreneurship education of colleges and universities with the national rural revitalization strategy more closely.It provides a feasible implementation path for higher vocational colleges to participate in the national rural revitalization strategy through innovative entrepreneurship education.

2 The great significance of the integration of innovation and entrepreneurship education in higher vocational colleges into the strategy of rural revitalization

With its unique way of running a school and clear objectives of personnel training, higher vocational colleges can accurately train advanced technical and skilled talents with all-round development of morality, intelligence, physique, beauty and labor for rural revitalization, by promoting innovative entrepreneurship education, better integration into the country's rural revitalization strategy is undoubtedly of great significance.

2.1 Promoting rural economic and social development

At the beginning of 2020, in the face of the surging COVID-19 epidemic, the CPC Central Committee led the people of the whole country to carry out an arduous fight against the epidemic, and the prevention and control of the epidemic achieved phased victories, ensuring the safety and health of the people; but to this end, China has also made a huge price, the economy and society are deeply affected by the epidemic, the task of poverty alleviation is more difficult, and the strategy of rural revitalization is facing greater challenges[1].

To carry out innovative and entrepreneurial education, mobilize teachers and students to participate in rural revitalization, develop rural e-commerce, and implement "Internet + agriculture" are of great significance in promoting rural economic and social development, solving the employment of rural labor force, and smoothly realizing the task of poverty alleviation and tackling key problems, and implementing the strategy of rural revitalization.

Driven by innovation and entrepreneurship education, we can not only solve the circulation problems of agricultural products, but also form channels conducive to expanding the market sales of agricultural products, so that farmers are no longer confused, so that agricultural products "have a place to go". It is easy to reduce the production risk of agricultural products, solve the problems of information asymmetry in the trading process of agricultural products, and solve the problems of smooth information flow of agricultural products in the market.

2.2 Improving the quality of running higher vocational colleges

In the new era, the development task faced by higher vocational colleges is more arduous. in the face of the task of high-level vocational colleges and high-level professional construction with Chinese characteristics, how to ensure the socialist school-running direction of the school? reflect the goal of the party and the state for vocational education, and effectively provide the necessary talents for economic and social development. In the coming 14th five-year Plan, more higher vocational colleges will enter the undergraduate school-running level.

Higher vocational colleges must make use of the rare opportunity of integrating innovation and entrepreneurship education into the rural rejuvenation to realize their own development and contribute to the great rejuvenation of the Chinese nation.

On the basis of running higher vocational colleges, we will continue to expand and extend the social service function through the strategy of rural revitalization, which will provide new driving force and support for the professional construction and personnel training of higher vocational colleges.

With the in-depth implementation of the rural revitalization strategy, especially under the background of the large-scale application of 5G technology, higher vocational colleges must speed up the mastery and application of new technologies and more actively serve the national development strategy [2]. Only by integrating the development and application of new technologies and new industries with innovative entrepreneurship education as a starting point, can professional construction be more dynamic and effective.

So as to further improve the school-running level and talent training ability of higher vocational colleges, and earnestly undertake the social responsibility of high-level higher vocational colleges with Chinese characteristics.

3 Opportunities and challenges of Innovation and Entrepreneurship in higher vocational colleges to serve rural revitalization strategy

Since 2020, COVID-19 has wreaked havoc, and the task of getting rid of poverty is arduous, and the strategic significance of rural revitalization is even more important. as the final year of the 13th five-year Plan, there are good opportunities for innovative entrepreneurial services in higher vocational colleges and the country's rural revitalization strategy. at the same time, it faces severe challenges.

3.1 The opportunities faced by college students in the process of connecting with the countryside through innovation and entrepreneurship

Standing in the historical period of the intersection of two century-old dreams, accurate poverty alleviation and rural revitalization are both important strategic arrangements of the CPC Central Committee and the solemn commitment of the Chinese Government to the people. Continuing to maintain vitality and creativity in the countryside has become the key to promoting rural revitalization, and college students are an important force to create vitality for rural development.

First of all, the release of preferential policies related to rural development and college students' innovation and entrepreneurship provides them with important opportunities. For example, in February 2019, the "opinions on the implementation of Agricultural and Rural work in 2019" issued by the Ministry of Agriculture and villages of the people's Republic of China clearly put forward the implementation of the "digital countryside" strategy and strengthening the innovation and promotion of agricultural science and technology. actively promote agriculture to go out, and the "opinions on further promoting the work of returning to hometown and starting a business" have further promoted the innovation and entrepreneurship of college students to connect with the countryside. Secondly, the arrival of 5G makes the Internet show more charm [3]. The mode of combining rural economic development with network science and technology can not only promote the effective integration of rural resources and improve the utilization rate of rural resources, but also provide a broad development space for college students to innovate and start their own businesses. Finally, the continuous enhancement of college students' awareness of innovation and entrepreneurship, and the continuous improvement of the school innovation and entrepreneurship platform, promote college students to actively move closer to the countryside in the process of innovation and entrepreneurship, provide more talents for the countryside, and promote the integration and development of rural industries.

3.2 The challenges faced by the students of higher Vocational Colleges in the process of connecting the Countryside with Innovation and Entrepreneurship

At present, although many students in higher vocational colleges have a strong enthusiasm for innovation and entrepreneurship, they can actively respond to the national call and strive to go deep into the grass-roots level, into the countryside and develop agriculture, but it is undeniable that most of the innovative and entrepreneurial projects of students in higher vocational colleges still remain at the creative level [4].

At the present stage, the bottleneck of natural resources appears in some rural areas and communities, which leads to the lack of corresponding innovative and entrepreneurial resources for students in higher vocational colleges when docking with the countryside; at the same time, the shortage of venture capital has become another obstacle for college students in the process of serving the countryside.

In addition to the obstacles brought by the rural areas, college students also have their own reasons for the problems existing in the service of rural revitalization in higher vocational colleges in recent years.

First of all, many college students do not have a deep understanding of the country's rural revitalization policy, and higher vocational college students are at an ambiguous stage of rural and community development and goal orientation. not to mention the work deployment of the national rural revitalization strategy and the characteristic policies of the corresponding areas [5]; secondly, the lack of social practice experience of higher vocational college students makes them at a disadvantage in raising resources and opening up markets [6]. Finally, due to the differences in ideas and considerations for their own future, the willingness of college students to go to the countryside is relatively low, which makes the countryside lack of corresponding human resources in the process of its development [7].

4 The concrete measures to realize the integration of innovation and entrepreneurship education into the strategy of rural revitalization

Innovation and entrepreneurship education focuses on the reform and innovation of talent training methods, and aims to serve the country's development strategy by breaking through the existing technology and mode of thinking. to realize the integration of innovation and entrepreneurship education into the rural revitalization strategy of serving the country should start from the following three directions.

4.1 To establish the family and national feelings of teachers and students in higher vocational colleges through innovative entrepreneurship education

The implementation of the rural revitalization strategy has a strong nature of the times. The education of innovation and entrepreneurship in higher vocational colleges will guide teachers and students to fully understand the country's rural revitalization development strategy and take root in the countryside. we should conscientiously carry out demand research, analyze the characteristics of rural agricultural products, understand the key points and difficulties of rural revitalization, pay attention to education and guide students to carry forward the spirit of labor, and combine "reading ten thousand books" with "traveling ten thousand miles".

Take root in China to understand the national conditions and people's conditions, increase wisdom and talent in practice, and temper will and character in the course of hard struggle.

At the same time, in the process of carrying out innovative and entrepreneurial projects, it can improve the scientific thinking methods of teachers and students, cultivate students' sense of responsibility and mission to explore the unknown, pursue the truth, and courageously scale the peak of science and technology. stimulate the team members' feelings of home and country and mission to serve the country with science and technology.

4.2 To guide teachers and students to promote the integration of new cross-disciplines in the process of innovation and entrepreneurship

As far as the specialty setting of higher vocational colleges is concerned, there are many professional fields related to rural revitalization, such as the theory of party building, grass-roots governance and other sociological fields, as well as VR, AR, Internet of things and other technologies extended by 5G new technology, as well as network technology, Web browsing technology, security technology, database technology, electronic payment technology, network data communication technology and other key technologies needed by e-commerce.

The application of new technologies will speed up the further integration of cross-disciplines, provide a good foundation for scientific and technological innovation, and enable the team members involved to better master and use these new technologies in the process of practice. cultivate the combination of science and education to cultivate innovative talents.
The team of innovative and entrepreneurial teachers and students who participate in rural revitalization should pay attention to strengthening the mastery and integration of new technologies in this process.

The team of innovative and entrepreneurial teachers and students who participate in rural revitalization should pay attention to strengthening the mastery and integration of new technologies in this process [8].

4.3 Give full play to the application and popularization of 5G and other new technologies, and promote the integration of college students' innovation and entrepreneurship education into rural revitalization.

Under the background of the rapid popularization and development of the Internet, the state has issued many policies to promote the coverage of rural e-commerce, and the "Government work report" proposes to implement the "Internet +" action plan. Among them, "Internet + Agriculture" is an important part of the plan.

E-commerce has become an indispensable means to implement the rural revitalization strategy. The "implementation Plan of Guangdong Province for promoting the Development of Rural E-commerce" released in 2016 mentions "vigorously supporting the construction of local e-commerce platforms for the management of famous agricultural products, agricultural products and rural tourism cultural products in the province".

E-commerce has come into people's field of vision, but also plays a more and more important role in people's lives, and has attracted the attention of many consumers.

The extensive use of 5G technology will also promote the further integration of innovation and entrepreneurship education and rural revitalization strategy.

From the perspective of the application of 5G technology, the research on the construction of rural e-commerce platform has strong innovation and practicability, and is of great significance for rural e-commerce how to serve the revitalization of rural areas. it has great potential for the integration and development of cross-fusion technology.

5 Acknowledgements

This work was financially supported by Scientific Research platform and projects of Guangdong Provincial Department of Education in 2020, Research on the Construction of Rural E-commerce platform under the background of 5G Technology (2020ZDZX1076).

References

1. Y. QU, Inner Mongolia Science Technology & Economy,**457**(15),10-11, (2020)
2. W. XU, Reformation & Strategy,**35**(5),55-62, (2019)
3. W. LUO, Journal of Wuhan Polytechnic University,**39**(2),73-79, (2020)
4. C.JIANG, Journal of Research on Education for Ethnic Minorities,**31**(3),26-30, (2020)
5. Y. ZHONG, Journal of Higher Education,**40**(8),88-97, (2019)
6. M Xu, H Wei, H Ni, China Higher Education Research,**257**(1),45-48,(2018)
7. X.LUO, Journal of Jiangsu Ocean University (Humanities & Social Sciences Edition),**17**(6),117-120, (2019)
8. X.WANG, Journal of Architectural Education in Institutions of Higher Learning, **28**(6),164-170, (2019)

Entrepreneurial Improvisation: Research Review, Integration Framework and Future Outlook

Xuejiao XU, *Yaqi* ZHAO*, and *Jing* ZHANG

School of Economics and Management, Beihua University, Jilin, China

Abstract. At present, the development context of entrepreneurial improvisation research is still unclear. This article reviews and sorts out the relevant literature on entrepreneurial improvisation from the individual and organizational levels: starting from the connotation, constituent elements and induction mechanism, classifies and sorts out the relevant research on entrepreneurial improvisation, and proposes an integrated framework and framework based on the summary of the frontier topics of entrepreneurial improvisation. The prospect of the future is expected to provide reference for the in-depth research and practice of entrepreneurial improvisation in the future.
Keywords: Entrepreneurial Improvisation, Improvisation Ability, Environmental Turbulence, Research Review

1 Introduction

Under the policy of "mass entrepreneurship and innovation", the whole people actively participate in the entrepreneurial boom. However, the sluggish success rate of entrepreneurship and the dynamic changes in the environment present provide great difficulties for entrepreneurs. How to make strategies that are most in line with the current and conducive to the future under the uncertain and unpredictable decision-making environment has become the key to the endurance of startups and a hot topic of academic discussion. Since Weick first introduced improvisation to the development of the management field, entrepreneurial improvisation has provided a new perspective for the survival and development of enterprises and is one of the research topics in the field of entrepreneurship.

Many literatures show that entrepreneurial improvisation can have a positive impact on the performance and sustainable development of enterprises. Impromptu greatly shortens the time gap between planning and execution, improves the effectiveness of resources at hand, and helps companies grasp the opportunity in a rapidly changing environment. In the process of business survival and growth, it is not an easy process to make appropriate plans. Start-ups need to consider improvisation, because it can change the strategic plan so that the company can respond immediately when opportunities arise (Arshad, Darwina, et al.,2018) [1]. Many scholars at home and abroad try to explain the formation mechanism of entrepreneurial improvisation from a certain theoretical perspective, but the research on the

* Corresponding author: zhaoyaqi9508@126.com

in-depth analysis of the development context of entrepreneurial improvisation is relatively scarce. Since entrepreneurial improvisation is still in its growth stage, systematically combing existing relevant literature is of great significance for the further research of entrepreneurial improvisation in the future. Based on this, this article reviews and sorts out relevant literature on entrepreneurial improvisation from the individual and organizational levels, hoping to provide a reference for in-depth research and practice of entrepreneurial improvisation in the future.

2 Research on the definition of entrepreneurship improvisation

2.1 Individual perspective

Weick (1993) first introduced organizational improvisation into the management field, and he regarded improvisation as a process-immediately inventing new methods and strategies to replace the old order [2]. Later, scholars introduced resource patchwork in the study of improvisation. Guoxiang Ruan et al. (2015) believed that employee improvisation is a behavioral way in which startups rely on employee spontaneity and creativity and use available resources to respond to and solve many unexpected technical and management problems in a dynamic environment [3]. On this basis, Ying Wang and Taoyong Su (2019) introduced emotion to the level of individual improvisation in subsequent empirical research related to improvisation, enriching individual level research, and defined employee improvisation as the creative use of resources at hand to quickly respond to task goals and solve accidents on the spot. Problem behavior is the result of comprehensive investment in cognition, emotion, and resources [4].

2.2 Organizational perspective

Moorman and Miner (1998) summarized organizational improvisation as "the degree of aggregation of creation and execution in time" [5]. Emphasis on time cohesion, but in subsequent research, spontaneity and innovation have become the new focus of the concept of organizational improvisation. Vera and Crossan (2005) are based on the process orientation of improvisation, and they believe that improvisation is a creative and spontaneous process to achieve organizational goals in a new way [6]. In addition, because the limited resources will restrict the development of the organization, Cunha et al. (1999) defined organizational improvisation as the concept of action carried out by the organization or its members based on available material, cognitive, emotional and social resources [7]. Leybourne and Sadler (2006) emphasized the influence of the organizational environment, and believed that organizational improvisation is driven by time pressure, a mixture of intuition, creativity and the use of available resources [8]. In addition, Zhuxin Ye et al. (2018) emphasized the contextualization of entrepreneurial impromptu, believing that entrepreneurial impromptu is the behavior of entrepreneurial teams re-integrating the resources available at their counterparts and consciously integrating the creation and execution of new solutions [9].

3 Research on the elements of entrepreneurship improvisation

3.1 Dimension Division

Early research on improvisation mainly focused on spontaneity, and the research structure was relatively simple. After that, the dimension of entrepreneurial improvisation was gradually expanded. Croosan and Sorrenti (1997) proposed to take intuition and spontaneity as the two dimensions of improvisation, and began to consider the influence of intuition [10]. Miner et al. (2001) believes that organizational improvisation includes four dimensions: novelty, consciousness, time aggregation and resource concentration [11]. On the basis of Vera, Hmieleski and Corbett (2006) divided improvisation into creativity and utilization of resources at hand, execution ability under high pressure environment, spontaneity and persistence [12], Jiang Yun and Chen Jiang (2010) inherited Miner's resource concentration Thought, it is proposed that organizational improvisation consists of three dimensions: immediate response, intention to create and use of existing resources [13]. In subsequent research, Lin Ding and Guangming Cao (2014) and others defined organizational improvisation at the organizational level into four dimensions: immediate response, intention creation, instant collaboration and resource integration [14].

3.2 Measurement method

Hmleleski and Corbett (2008) et al. selected 12 of the above 27 items as a more comprehensive scale for measuring individual improvisation [15]. Massimo (2009) et al. modified the scale on the basis of the scale developed by Vera and Crossan (2005), using a seven-item scale to measure the degree of individual improvisation, which mainly expresses the two aspects of creativity and immediacy [16].Moorman and Miner (1998) used two methods to measure improvisation at the same level of analysis relative to planning: the time difference between plan and execution, and the degree of deviation from the previous plan [5], Subsequently, in his research on the improvisational behavior of new product development teams, the scale was used for the first time to measure improvisational behavior, which laid the foundation for the development of improvisational scales. Vera and Crossan (2005) developed a relatively complete organizational improvisation scale on the basis of the existing employee creation power scale, spontaneity measurement and improvisation measurement, with a total of 7 tests [6]. Lin Ding and GuangMing Cao (2014) developed an organization-level improvisation scale based on the research on the organizational improvisation ability of commercial banks, which specifically includes 4 dimensional variables and 19 measurement items [14].

4 Research on the induction mechanism of entrepreneurship

4.1 Management related factors

In previous studies, leadership style is one of the important influencing factors of improvisational behavior. For example, the essence of transformational leadership helps employees not be restricted by the organizational framework, and while supporting employees, it also makes them improvise in necessary situations (Haidong Li and Zhiyang Lin ,2013) [17]. Similarly, inclusive leadership uses a leadership style that respects the personality of employees, encourages employees to play positively and tolerate mistakes, enhances employee recognition and sense of belonging, and promotes employee improvisation. Entrepreneurial leadership creates an atmosphere of innovation for the organization, giving people self-confidence. Under the call of innovation, employees can have an improvisation that is inherently compatible with entrepreneurial leadership (Ying Wang and Taoyong Su ,2019) [4].

4.2 Emotional social interaction related Factors

Research has found that organizational emotions triggered by contextual changes can provide an information source for the organization's temporary performance, generate instant activities suitable for the current situation, and can be adjusted in real time based on emotional responses. Organizational impromptu behavior is affected by the relationship between organizational emotional response and organizational action The positive impact of convergence. In addition, some scholars believe that relationships can help some companies obtain valuable resources and information, thereby promoting the sustainable development of companies. At the same time, the internal network of the enterprise is also conducive to the improvisation of employees.

4.3 Resource related factors

Improvisation depends on the resources at hand, and the implementation of creative ideas requires sufficient resource support. Therefore, the availability of resources is conducive to the use of individual improvisation capabilities (Haidong Li and Zhiyang Lin,2013) [17]. In addition, organizational memory is a knowledge resource, which can provide rich experience and materials to help the organization quickly adapt to changes in the current situation, and the higher the level of organizational memory, the greater the possibility of improvisation. Some scholars questioned this. Studies have shown that long-term use of organizational memory for immediate use can easily lead to organizational rigidity and machinery, which is not conducive to organizational improvisation.

4.4 Environmental atmosphere related Factors

In order to ensure the survival and development of enterprises, organizations need to construct new behaviors immediately to face the changed environment. As far as the internal environment of the organization is concerned, when the organization advocates an error management atmosphere that tolerates errors and learns from experience in time, it is conducive to individuals seeking and obtaining valuable information and higher creativity, thereby increasing the probability of improvisation in the organization. In addition, the higher the level of team behavior integration and the greater the cohesion of the team, the higher the degree of improvisation shown by individual members (Leybourne and Sadler ,2006) [8].

5 Integration framework construction and future prospects

5.1 Integration framework construction

On the basis of combing the existing literature, this study constructed an integrated framework model (refer with: Fig.1)

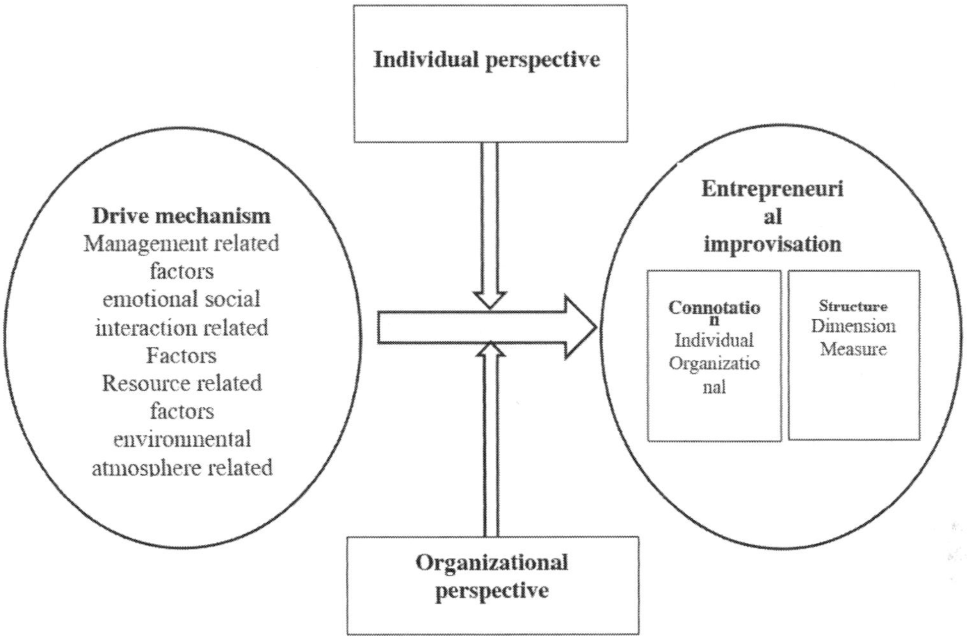

Fig. 1. Integrated research framework.

5.2 Future research outlook

The changing and unpredictable nature of today's business environment has made improvisation an increasingly common phenomenon in the entrepreneurial field. Although the research on entrepreneurial improvisation has made great progress, it is still in the stage of continuous exploration. Future research needs to further deepen the following aspects: First, improve the conceptual connotation of entrepreneurial improvisation and standardize the field of research. The connotation of the concept of improvisation is not the same, so many scholars have to distinguish it from other confusing concepts, which is not conducive to the research process of entrepreneurial improvisation. Under the current contextualization of entrepreneurship, an accurate definition of entrepreneurial impromptu is urgently needed, so as to explore the components of improvisation in depth and provide clear guidance for further specific research. Second, promote the output of empirical research. Pay attention to the inducing mechani21e Chinese context in order to explore the inherent characteristics of entrepreneurial improvisation.

References

1. D. Arshad, N. Zakaria, K. Abdul-Kadir, N. Ahmad, Linkage between Flexibility and SMEs Performance: Does Improvisation Matter? In MATEC Web of Conferences, **215**(2018)
2. K.E. Weick, The Collapse of Sense making in Organizations: The Mann Gulch Disaster. Administrative Science Quarterly,**38**(4),628-652, (1993)

3. G X Ruan, J Q Mao, L J Ma, A Cross-Level Research on the Mechanism of Employee Impromptu Behavior on Individual Innovation Performance——Based on the Empirical Study of New Energy Startups, China Soft Science, (1),108-117, (2015)
4. Y Wang, T Y Su. A Study on the Impact Mechanism of Entrepreneurial Leadership on Employee Impromptu.Soft Science,**33**(11),69-74, (2019)
5. C.Moorman, A.S. Miner, Organizational Improvisation and Organizational Memory, Academy of Management Review,**23**(4),698-723, (1998)
6. D.Vera, M. Crossan, Improvisation and innovative performance in teams, Organization Science,**16**(3),203—224(2005)
7. M. P. Cunha, J V Cunha, K Kamoche , Organizational Improvisation: What, When, How, and Why,International Journal of Management Review, **1**(3),299-341,(1991)
8. S Leybourne, E Sadler-Smith. The Role of Intuition and Improvisation in Project Management, International Journal of Pro- ject Management, **24**(6), 483-492, (2006)
9. Z X Ye, Y Y Mai, L Y Wang. The Status Quo and Future Prospects of Research on Impromptu Behavior of Startups, Foreign Economics and Management ,**40**(4),16-29, (2018)
10. M Croaan, M Sorrenti, Making sense of improvisation.Advances in Strategic Management,**14**(3),155-180,(1997)
11. A S Miner, P Bassoff, C Moorman, Organizational improvisation and learning: A field study, Administrative Science Quarterly, **46**(2),304 - 337, (2001)
12. K.M. Hmieleski, A. C. Corbett. Proclivity for Improvisation as a Predictor of Entrepreneurial Intentions, Journal of Small Business Management ,**44**(1), 45-63, (2010)
13. J Yun, C Jiang, A Review of Research on Foreign Organizational Improvisation, Foreign Social Sciences, (5),4-12, (2010)
14. L Ding, G M Cao. The Dimensional Construction and Measurement of Organizational Impromptu--Based on the Empirical Research of Chinese Commercial Banks, East China Economic Management ,**000**(009),84-89, (2014)
15. K. M. Hmieleski, A C. Corbett. The contrasting interaction effects of improvisational behavior with entrepreneurial self-efficacy on new venture performance and entrepreneur work satisfaction. Journal of Business Venturing, **23**(4),482–496, (2008)
16. M. Magni, L. Proserpio, M. Hoegl, B. Provera. The role of team behavioral integration and cohesion in shaping individual improvisation, Research Policy, **38**(6), 1044-1053, (2009)
17. H D Li, Z Y Lin. Study on the Antecedent Variables of Employees' Impromptu Behavior, Journal of Jiangxi University of Finance and Economics, (03),25-36, (2013)

Analysis on Infectious Risk Stress in the Supply Chain of Mega Infrastructure Project

Na Zhao*

School of Finance, Harbin University of Commerce, Harbin 150001, China

Abstract. In the implementation process of great infrastructure engineering, the structural safety of supply chain in coping with infectious risks (IR) is of important significance to operation of the whole project. Moreover, infectious risk can influence safety of supply chain of mega project (MP). Therefore, this study constructed a function of infectious risk stress, calculated the infectious risk stress value of supply chain of mega infrastructure project (MPSC), and analysed influences of random variables on infectious risk. Results demonstrate that the maximum infectious risk stress is at the end of supply chain of mega infrastructure project. Research conclusions provide references to anti-risk integrated design of supply chain of mega infrastructure project.

Keywords: Supply Chain, Infectious Risk, Mega Infrastructure Project

1 Introduction

The global economy is facing with great challenges due to influences of epidemics. Accelerating infrastructure layout and construction is the key to promote economic recovery. Due to the adding of new infrastructure project (NIP), infrastructure project is vested with more connotations and covers a wider scope at present. NIP mainly focuses on 7 fields, including 5G infrastructure construction, extra-high voltage and intercity high-speed railways, intercity rail traffic, charging piles for new energy automobiles, big data centre, artificial intelligence (AI) and industrial Internet. NIP involves additional economic digital characteristics compared to traditional infrastructure project centred at steel and cement. As an integrated management idea and method, supply chain management of mega projects is a management process that integrates planning, organization, coordination and control over logistics, information flow, capital flow, value-added flow, business flow and relations of other subjects in the supply chain [1]. In this environment, it requires not only coordination of internal links like planning, purchasing, manufacturing and marketing, but also close cooperation between upstream and downstream subjects including suppliers and distributors [2-4].

Supply chain of mega projects has real-time impacts on construction practices of mega infrastructure project and structural safety of the supply chain has important impacts on safe operation of the whole project, while some certain and uncertain infectious risks influence safety of supply chain of mega infrastructure project [5]. The infectious risk of mega

* Corresponding author: zhaona@hit.edu.cn

infrastructure project refers to the risk posed by the anti-social behaviour, counterproductive behaviour, altruistic behaviour, and conflicts of a certain group or individuals or other stakeholders of the supply chain of MP of the impacts of external and internal factors [6-8]. Researches on infectious risk stress of supply chain mainly concentrate on risk recognition, assessment and combined strategy. Nevertheless, specific calculation of infectious risk stress in a supply chain of mega projects have to be supplemented and perfected urgently [9,10]. Based on review of Chinese and foreign researches, this study attempted to calculate infectious risk stress of supply chain of mega projects, and quantized the infectious risk stress value by combining risk certainty. Additionally, the action mechanism of infectious risk pressure was disclosed by analysing influences of random variables on infectious risks. Research conclusions provide references to anti-risk integrated design of supply chain of mega infrastructure project and to improve reliability as well as security [11].

2 Calculation methods and procedures

Supply chain of mega projects is a system composed of many subjects, including investors, builders, contractors, suppliers, forwarders, etc. Supply chain management is to plan, coordinate, operate, control and optimize various activities and processes in the whole supply chain system and it is to assure completion of mega projects on schedule according to accurate specification, quality and state at the lowest total cost. Obviously, MP supply chain (MPSC) management is a management mode which reflects the idea of integration and coordination. It requires cooperation of members in the MPSC system in coping with complicated and changing external environment [12].

To address this problem, the infectious risk stress of supply chain of mega infrastructure project was calculated by random sampling method. The calculated results approached to accurate solutions gradually with the increase of simulation times. The basic method is to generate a group of special random variable values artificially in the corresponding probability distribution by using an algorithm, and then implement abundant simulations, and finally estimate risks by testing simulation results. Accuracy of random sampling method is determined by simulation times. The key of random sampling method is to generate a lot of random numbers in known distribution and then implement many simulations based on random numbers in a short period. Specific steps are introduced as follows: (1) Setting up the function D and analysing probability distribution of various random variables. (2) Bringing random numbers which obey to probability distribution into the function D. (3) Making statistics on times t with $D < 0$. (4) Calculating the risk rate $Q_{\mathrm{f}} = \frac{t}{T}$, where T is the total number of simulations.

3 Function for infectious risk stress calculation of MPSC

Basic variables which influence the structural state of MPSC are $a_1, a_2, \cdots, a_t$, and they can be classified into two basic variables according to attributes, namely random variable P and random variable W of risk stress strength. Therefore, it can get:

$$\begin{gathered} P = P\left(a_{P_1}, a_{P_2}, \cdots, a_{P_i}\right) \\ W = W\left(a_{W_1}, a_{W_2}, \cdots, a_{W_j}\right) \end{gathered} \tag{1}$$

where a_{P_i} is a variable related with structural intensity of MPSC. a_{W_j} is the variable related with risks, including investors, builders, contractors, suppliers and forwarders. Based on such processing, the problem of multiple random variables is transformed into a problem of two random variables. For the convenience of analysis, the function of MPSC structure can be simplified as:

$$D = g\ (P,\ W) \tag{2}$$

For Eq. (2), different forms can be chosen according to different situations. The function of infectious risk stress of MPSC is determined:

$$D = P - W \tag{3}$$

Here, the random variable P of infectious risk stress of MPSC is composed of $P = P\ (\mu_W,\ \gamma)$, where μ_W is the compromise strength of suppliers and γ is the collaborative coefficient. For the random variable W of infectious risk stress of MPSC, there's $W = \mu_\mathrm{p}$ since the MPSC is in the multiple influence state. According to characteristics of infectious risk of MPSC structure, the function is designed as follows:

$$D = P - W = \gamma\mu_W - \mu_\mathrm{p} \tag{4}$$

where μ_p is the internal risk of supply chain of MPSC.

The random variable of infectious risk stress of MPSC structure is $W = W\ (\mu_a,\ \mu_\mathrm{p})$ and the relevant function is built up:

$$D = -\sqrt{\mu_\mathrm{p}^2 + \mu_\mathrm{a}^2 - \mu_\mathrm{p}\mu_a} + \gamma\mu_W \tag{5}$$

where μ_a is an external risk.

4 Distribution characteristics and statistical eigenvalues of random variables

In addition to major random variables that influence structural safety of MPSC, this study also viewed some parameters as random variables. Among them, statistical eigenvalues of planning coefficient γ, collaboration L_j and operation L_k can be gained through relative quantity:

$$\left|\begin{array}{l} \vartheta_j = L_{\max}/L_\mathrm{d} \\ \vartheta_k = L_k/L_\mathrm{wl} \\ \vartheta_\gamma = \gamma_W/\gamma_\mathrm{g} \end{array}\right. \tag{6}$$

where $L_{\max}$ is the maximum risk of MPSC, L_d is the design risks, L_k is the central operation condition of MPSC, L_wl refers to control, and γ_W and γ_g are coefficients.

There are many influencing factors of MPSC risks and there's no relevant data for mathematical statistical analysis. According to error theory and central-limit theorem, these influencing factors were viewed belonging to normal distribution.

5 Conclusions

Based on function calculation, the maximum infectious risk stress is observed at the end of supply chain of MPSC structure. Therefore, the end of infectious risk stress of MPSC is used as the marking point in risk calculation and analysis. With references to probability distribution and statistical eigenvalues of different simulation variables, it can be seen from random sampling method that selection of standard deviations for different simulation variables can influence calculation of infectious risk stress of MPSC structure significantly.

To disclose influences of a simulation variable on comprehensive infectious risk stress of MPSC structure, the standard deviation of one variable is changed independently and uncertainty of infectious risk stress is the primary influencing factors of risks. To sum up, risk rate increases with the increase of the ratio between practical value and design value, indicating that uncertainties of investors, builders, contractors, suppliers and forwarders can improve stress resistance of MPSC structure slightly. Moreover, inherent feature parameters of mega projects can be viewed as certainty variables. Key attentions shall be paid to improvement of stress resistance of the MPSC under infectious risks through planning, coordination, operation, control and optimization.

Acknowledgments

This research was supported in part by the Natural Science Foundation of Heilongjiang of China under grant number QC2017084, and Heilongjiang Postdoctoral Foundation of China.

References

1. G D Wu, *Project-Based Supply Chain Cooperative Incentive Based on Reciprocity Preference*. International Journal of Simulation Modelling,**13**(1), 102-115 (2014)
2. Y J Zheng, H F Ling, *Emergency transportation planning in disaster relief supply chain management: a cooperative fuzzy optimization approach*. Soft Computing, **13**(1), pp 1301-1314(2013)
3. I Karlsson; J M Rootzen, F Johnsson, *Reaching net-zero carbon emissions in construction supply chains - Analysis of a Swedish road construction project*. Renewable & Sustainable Energy Reviews, **120** (2020)
4. M Staniek, G Sierpiński. *Smart platform for support issues at the first and last mile in the supply chain - the concept of the s-mile project*. Scientific Journal of Silesian University of Technology. Series Transport, **92**, 141-148 (2016)
5. B Marcus, G Kannan, S Joseph et al. *Quantitative models for sustainable supply chain management: Developments and directions*. European Journal of Operational Research, 299–312 (2014)
6. M Bostrom, *A missing pillar? Challenges in theorizing and practicing social sustainability*, Sustain Sci Pract Policy, 3–14 (2012)
7. B Stephen; M Andrew. *Does it pay to be different? An analysis of the relationship between corporate social and financial performance*. Strategic management journal. **12**(27), 1325-1343 (2008)
8. R A Buchholz, S B Rosenthal, *Toward a contemporary conceptual framework for stakeholder theory*, Journal of Business Ethics, **58** 137-148 (2005)
9. C R Carter, M.M Jennings, Logistics social responsibility: an integrative framework. *Journal of Business Logistic*, **23**(1), 145-180 (2002)
10. C R Carter, D S Rogers. *A framework of sustainable supply chain management: moving toward new theory*. International Journal of Physical Distribution & Logistics Management, **38**(1), 360-387 (2008)
11. F Zair, N Sefiani, M Fourka. *Advanced optimization model of resource allocation in the B2C supply chain*. Engineering review (2018)
12. C Haiwook, C H Yeon. *A Model for Interorganizational Information Technology Infrastructure and Electronic Cooperation Between Firms in Supply Chain*. AMCIS Proceedings **318** 2002,

An Analysis of the Obstacles to the Implementation of BIM in China's Project Cost

Lei Tian, *Qiong* Shen* and *Yanan* Ma

College of Architecture and Urban-Rural Planning, Sichuan Agricultural University, Dujiangyan Sichuan, 611830, China

Abstract. BIM has huge potential in the construction industry, but the development in the field of project cost is not as good as expected. In this paper, the factors that prevention BIM implemented in project cost were sorted out, and the factors were graded by questionnaires, then factor analysis method and importance index were used to calculate the biggest influence factor of BIM in project cost. This paper puts forward some measures from the perspectives of personnel, software, cost and enterprises, such as strengthening cooperation between universities and enterprises to cultivate fine talents, increasing scientific research investment to perfect BIM cost software, enriching BIM learning resources to increase the study enthusiasm of cost workers, and establishing reasonable cost-sharing mode. In order to promote the development of project cost information in China.
Keywords: BIM, Project cost, Factor analysis, Obstacle factors

1 Introduction

Along with the progress of the whole construction industry, the project cost information has got certain development. The latest technologies have provided more possibility for the construction industry's benign development, among which BIM technology is particularly favored by everyone. It can be used in design, construction, management and so on. It has the advantages of visualization, coordination, simulation, optimization, integration, parameterization, information completeness and map-making. With the increasing demand for building quality, there is an increasing demand for BIM to be applied to building. Therefore, BIM3D and BIM4D technology were born. 4D is based on 3D technology, adding time management, and can control the progress of the project. Then, the BIM5D technology, which is beneficial to cost management, can be used to simulate the construction process by combining the building model, schedule and cost management with the BIM platform, and provide contract management information. So that the project can be in all stages of information sharing, saving time and cost.

Most researches on BIM in China remain at the analysis stage, lacking systematization and sufficient experience to support it. In order to implement BIM, the government has also formulated relevant policies. In August 2017, *Ministry of Housing and Urban-rural Development* issued the "13th five-year Plan" of the project cost, which proposes to

* Corresponding author: swydong@sicau.edu.cn

vigorously promote the application of BIM technology in project cost. However, there is still a long way to go to achieve the goal of integrating BIM with management systems and other information technologies. In order to speed up this process and realize the popularization of BIM in project cost as soon as possible, some scholars have studied the difficulties faced by the implementation of BIM. Gao et al. [1], Xu et al. [2] respectively used factor analysis, DEMATEL analysis and ISM analysis to study the influencing factors of the application of BIM in China's construction industry and corresponding countermeasures and suggestions were put forward. It provided reference for perfecting BIM application standard and promoting the popularization of BIM in China. Zheng et al. [3] indicated that enterprises should pay attention to both technical upgrading and teamwork to promote the application of BIM. Liu et al. [4], Xue et al. [5] focused on the cost process of BIM application barriers and the main problems in the proposed measures. so that the project cost can realize informatization as soon as possible. Zhu et al. [6] discussed the obstacles and system guarantee system of project cost information construction in order to put forward feasible suggestions for the reform of project cost consultation industry. The previous researches lack of BIM in the project cost implementation of barriers to explore the relationship between factors. So, this paper used factor analysis and importance index to calculate the degree of influence of the obstacles. Finally, some suggestions are given to solve the problem.

2 Obstacles to the development of BIM in the cost industry

In this paper, the existing adverse phenomena and relevant literature were analyzed, then some representative obstacles were sorted out. The obstacles were made into an expert questionnaire, and the personnel who have been engaged in relevant research or work for more than 3 years are consulted. The result is a list of 18 obstacles of BIM in the project cost, as Table 1. Factor analysis method and importance index were used to analyze the obstacles proposed and explore the relationship between them. The results obtained can provide targeted recommendations for the implementation of BIM in the project cost.

Table1. Obstacles to the implementation of BIM in the project cost.

Code	Influence factor	Common factorV1	Common factorV2	Common factorV3	Common factorV4
F1	BIM lack standard system	0.165	0.226	**0.695**	0.284
F2	BIM in the project cost lack overall management	**0.581**	0.508	0.23	0.025
F3	BIM in the project cost raise industry requirements	**0.692**	0.148	0.219	-0.088
F4	The publicity work of promoting BIM application in the project cost is not in place	**0.787**	0.252	0.032	0.09
F5	The application of BIM in the cost industry lacks professional talents	**0.743**	0.111	0.191	0.188
F6	The application of BIM in the project cost government support is insufficient	0.402	0.299	**0.503**	-0.289

F7	The application of BIM in the project cost faces negative treatment from enterprises	**0.572**	-0.134	0.315	0.514
F8	Software purchases and hardware upgrades are expensive	**0.733**	0.123	0.097	0.338
F9	The project cost industry information sharing is not timely	0.412	**0.627**	0.278	0.014
F10	BIM cases and practical experience are relatively few	0.378	**0.525**	0.336	0.15
F11	The industrial dispute processing mechanism is not perfect	0.151	**0.841**	0.044	0.067
F12	The synergy ability of the industry needs to be improved	0.058	**0.759**	0.393	0.059
F13	BIM software is not fully compatible with cost software	0.119	0.524	0.12	**0.607**
F14	The work flow of the project cost restricts the application of BIM	0.158	0.257	**0.774**	0.156
F15	The BIM models have unclear intellectual property rights in its cost application	0.03	**0.802**	0.244	-0.235
F16	The economic benefits brought by BIM are not significant enough	0.342	-0.03	0.195	**0.702**
F17	The cost of personnel training is high	**0.743**	0.054	0.03	0.395
F18	Lack of testing methods for accuracy of BIM models for cost work	0.123	**0.718**	-0.019	0.471

3 Research methods

3.1 Questionnaires design and distribution

The survey was designed using a 5-level Likert scale to rate the obstacles to BIM implementation in terms of project cost. Questionnaires distributed in the range of Chengdu and Chongqing, two cities, and the issuance of the questionnaires cover the design, construction, and real estate industry. The issuance of the questionnaires mainly adopts two ways: through the individual distributed, construction industry workers and construction cost workers were investigated; Through the instructor to carry out the questionnaires, most of the respondents served as a project manager or technical director of the project. A total of 120 questionnaires were issued in this survey, and 98 valid questionnaires were recovered, with the effective questionnaires recovery rate of 81.7%. Among them, there are 69 questionnaires from relevant employees and 29 questionnaires from graduate students and tutors of relevant majors in universities.

3.2 Data processing

Questionnaire reliability test. Reliability is to test the validity of statistical data, and Cronbach's A is generally used as the index of reliability test. Usually, Cranbach's A above 0.6 indicates reliable statistical data. In this paper, the coefficient of Cranbach's A is 0.908 through Spss25.0 calculation, indicating that the survey data credibility of the questionnaires is good.

To extract the key factors. The significance coefficient of KMO value and spherical test is used to judge whether the original factor indicator variable is suitable for factor analysis. If the KMO value is greater than 0.5, it indicates that there is a high correlation between the original factor indicator variables and it is suitable for factor analysis. In sphericity test, if the significance coefficient is less than 0.05, it is appropriate to carry out factor analysis on the original factor indicator variable. According to Spss25.0 calculation in this study, KMO value was 0.801 greater than 0.5, and Sig value was 0.000, less than 0.05. So the survey data were suitable for factor analysis. In this study, principal component analysis was used for factor analysis of original factor variables, and factors with eigenvalues greater than or equal to 1 were selected as key factors. Through calculation, 4 common factors were extracted, and the cumulative variance contribution rate was 67.067%, which was more than 60%, indicating that these four factors could summarize most of the information of the 18 factors. At the same time, orthogonal rotation of the factor component matrix was performed by the maximum variance rotation method. The rotated factor component matrix was shown in Table 1. The maximum factor loading value of each initial variable relative to each principal component is found in the rotated factor component matrix (bold numbers in Table 1) and merged into the common factor.

As Table 1, the 18 questionnaire indicators are divided into 4 common components. Common component V1 include: F2、F3、F4、F5、F7、F8、F17, and it was named as industry influence factors. Common component V2 include: F9、F10、F11、F12、F15、F18, and it was named as environmental influence factors. The common component V3 includes: F1 、F6 、F14, and it was named as system influencing factors. Common component V4 includes: F13、F16, and it was named as software influencing factors.

3.3 The importance of calculating key obstacle factors

The importance index of key obstacle factors is calculated, using the following formula:

$$\frac{(a_1m_1 + a_2m_2 + K + a_nm_n)}{98} \times \frac{100}{5} \tag{1}$$

a_n is in the score of the nth obstacle factor index; m_n is the number of questionnaires with the same score for the nth obstacle index. The calculation results are shown in Table 2.

Table 2. Importance calculation table.

Module	Common factor	Importance index	Ranking	Obstacle factor	Importance index	Ranking
industry influencing factors	V1	81.14	1	F2	74.29	10
				F3	74.49	9
				F4	74.29	10
				F5	89.39	1
				F7	83.67	4
				F8	87.35	2
				F17	84.49	3
environmental influencing factors	V2	67.21	4	F9	74.08	12
				F10	77.96	7
				F11	61.43	17
				F12	62.86	16
				F15	59.8	18
				F18	67.14	15
system influencing factors	V3	74.56	3	F1	77.14	8
				F6	72.86	14
				F14	73.67	13
software influencing factors	V4	79.9	2	F13	80	5
				F16	79.8	6

From Table 2, industry influence factors play an important role in the promotion of BIM in the project cost, among which the lack of professional talents is the biggest obstacle for BIM to be used in the project cost. The high cost of software purchases, hardware upgrade and personnel training and the application of BIM in the project cost faces negative treatment from enterprises also have great impact. To promote the popularization of BIM in the project cost, we need to solve the outstanding problems, and then gradually promote the in-depth application of BIM in the project cost.

4 Suggestions

Through the above researches, in order to better develop BIM in the project cost, the following suggestions are put forward:

(1) Paying attention to the cultivation of fine talents. The shortage of professional refined talents is an urgent problem to be solved at present, which requires the cost talents not only to master the specifications of engineering quantity calculation, but also to be skilled in modeling software. Furthermore, they also need to know how to use resources to reduce costs and complete real-time dynamic cost management to achieve process control. For the cultivation of the talents can adopt the way of cooperation between colleges and enterprises. The school adjusts the student development plan according to the needs of the enterprise. Enterprises provide more resources and technology for schools, which can help schools deepen education. Through this way, the software operation ability and practical application

ability of the relevant employees can be improved, and the whole process of cost management can be refined.

(2) Increase the investment in BIM research and improving the functions of cost software. At present, most of the BIM software in our country's cost industry is imported from abroad, and the statistical rules of software embedding in foreign countries are not completely unified with the rules of domestic engineering quantity calculation, therefore, it is inevitable that due to the applicability problem, the software functions are missing or the application effect is not ideal. The introduction and maintenance of BIM related cost software as well as the hardware upgrades cost of enterprises are all relatively high, which lead to the low popularity of BIM software application in the domestic project cost consulting industry. Therefore, to promote BIM in the cost industry, the research and development department needs to localize BIM software, improve software compatibility, reduce software requirements, and reduce software costs. In order to promote the application of BIM software in the cost industry and improve the information level of the cost industry.

(3) Enriching BIM learning resources and enhancing the enthusiasm of cost personnel in independent learning. At present, there is a shortage of BIM cost personnel with strong learning ability and rich practical experience in China, and even less BIM experts who can provide training and guidance for project cost personnel, which invisibly increases the difficulty of BIM learning. In addition to spending a lot of time on software operation training, cost practitioners should accumulate experience through constant practice to become mature BIM cost personnel. In order to solve this problem, we need to lower the learning threshold of BIM, enrich the learning resources, organize the exchange of experience and discussion of BIM experts and technical staff at home and abroad, and formulate incentive policies. With the increase of the initiative of the cost personnel, the time and cost of training BIM cost personnel will be reduced.

(4) Improving the enthusiasm of the enterprise. Lack of BIM awareness among senior management is a common phenomenon in the current construction industry, and the cost and benefit of BIM application do not match for the enterprise. Both discourage managers from applying BIM. In the case of immature market rules and imperfect guidance standards, the owner holds a wait-and-see attitude towards the application of BIM in cost management by the cost consulting industry. Therefore, the government's formulation of incentive policies to provide guarantee is the basis for promoting the promotion of BIM in the construction cost industry in China. On this basis, the issue of cost distribution should also be considered. Cost distribution should be carried out according to all interested parties, to reduce the concerns of owners and enterprises and reduce obstacles for the implementation of BIM in the construction cost industry.

5 Conclusion

In this study, questionnaire survey and factor analysis were used to extract 4 common influencing factors of BIM application in the project cost. This paper puts forward some measures from the perspectives of personnel, software, cost and enterprises, such as strengthening cooperation between universities and enterprises to cultivate fine talents, increasing scientific research investment to perfect BIM cost software, enriching BIM learning resources to increase the study enthusiasm of cost workers, and establishing reasonable cost-sharing mode. They provide solutions to the outstanding problems that BIM faces in the development and popularization of the project cost in China, and promote the development of project cost information in China.

References

1. M. Gao, S.Y. Hao, A Study of Influencing Factors for BIM Application in Chinese Construction Industry Based on Factor Analysis, J Engin Mgt, **33**(4), 38-42(2019).
2. Y.Q. Xu, Y.Y. Kong, Analysis of the Influence Factors of Application and Promotion of BIM in China, J Engin Mgt, **30**(2), 28-32 (2016).
3. Y.J. Zheng, X. Ma, Y.Z. Su, T. Li, Y. Zhang, P. Lu, Research on Barriers of BIM Application in Construction Enterprises—Based on Literature Review Method, J Info Tech Civ Engin Arch, **11**(5), 61-65(2019).
4. C. Liu, J. J. Guo, Study on Barriers of Building Information Modeling in the Process of Cost, J Engin Mgt, **30**(2), 33-38(2016).
5. D. Xue, Research on application barriers of BIM in the cost process, Tech Econ Gui, **26**(3)206-208(2018).
6. X.S. Zhu, J.P. Peng, Barriers and Institutional Security System of China's Project Cost Information Construction, J Engin Mgt, **29**(2), 22-26(2015).

R&D Investment and Patent Application: The Effects of Internationalization and Ownership

Francis Boadu[1,2], *Hongjuan* Tang[3] and *Yu* Xie[4] *

[1]School of Management and Economics, University of Electronic Science and Technology of China, Chengdu 611731, China
[2]Faculty of Entrepreneurship and Enterprise Development, Kumasi Technical University, Kumasi, Ghana
[3]School of economics and management, chongqing jiaotong university, Chongqing 400067, China
[4]School of Management, Chongqing Technology and Business University, Chongqing 400067, China

Abstract. The purpose of this paper is to assess whether research and development (R&D) investment, degree of internationalization (DOI), and ownership structure affect the patent application of enterprises in an emerging market. Using a firm-level data from 242 China's information and communication technology (ICT) and automobile manufacturing listed companies on Shanghai and Shenzhen A- stock exchanges market from 2011 to 2015, our empirical results demonstrate that: (1) the degree of internationalization of enterprises has a positive facilitating influence on the correlation among corporate R&D investment and a patent application; (2) the state-owned ownership negatively moderates the R&D investment-patent application link, and (3) international expansion can adequately compensate for the efficiency disadvantages of R&D investment and patent application engender by state ownership. Besides, the study discovered and explained an exciting phenomenon and bring forth important enlightenment to enterprises in the emerging economies to improve resource utilization efficiency and achieve technical ability to catch up.
Keywords: R&D investment, Innovation, Internationalization, Ownership; emerging economies

1 Introduction

Technological innovation is regarded as a critical driving element to enhance enterprises' sustainable competitive advantage, productivity, and growth [1]. If used appropriately, it can aid enterprises to become more successful and sustainable in today's competitive environment. For instance, most global giants or enterprises from advanced economies have benefited from technological innovation due to their consistent effort to pursue technological breakthroughs to address the consumption patterns of both citizenry and customers from the international arena. However, the situation may be quite different in emerging economies, especially in China. Intriguingly, the proponents of resource dependence believe that Chinese enterprises are backward in terms of technology, and therefore have disadvantages in the

* Corresponding author: xieyu@ctbu.edu.cn

global market environment. Indeed, for Chinese enterprises to address and erase this perception towards achieving post-development technology, it is prudent for the enterprises (i.e., state or private) to improve the research and development (R&D) resource utilization efficiency. In recent years, this phenomenon has led scholars to focus more on how the internal and external factors such as R&D resource redundancy, foreign investment behavior, international expansion, absorptive capacity, ownership structure, institutional resources, managerial corruption, and governmental intervention of enterprises in emerging economies affect R&D investment and innovation output towards sustainable development. These, prior studies from the extant literature, have exposed some of the factors affecting the R&D investment-output relationship in the emerging economies.

However, there are still shortcomings in the current economic globalization that need further probing to unravel the exact impact among the variables. First, the above studies conducted by scholars focus on the internal governance of the enterprises, with scant attention paid to the variation of corporate innovation practices caused by the cross-border expansion of China's local "latecomers" in the context of resource dependence. Secondly, ownership and internationalization, as important factors affecting the positive correlation amongst R&D investment and a patent application of local Chinese enterprises under the background of economic globalization, are often studied separately and lack the exploration of the interaction mechanism. Considerably, we argue that the complex interaction of these variables (i.e., ownership and internationalization) can play a vital and stimulating role in addressing the gap in prior research. Hence, the current study intends to explore the combined influences of ownership and internationalization on the R&D investment-patent application relationship, in the following conditions: (i) the global competitive environment, and (ii) the change under different internationalization levels. Specifically, it seeks to answer the following: RQ1. Is the correlation among R&D investment and patent application dependent on an enterprise's involvement in international expansion? RQ2. Does state-ownership negatively moderate the correlation between R&D investment and patent application? RQ3. Does international expansion have a positive influence on the negative effects of state-ownership on R&D investment and patent application? Using firm-level data obtained from China's ICT and automobile manufacturing companies listed on Shanghai and Shenzhen A-stock exchanges market from 2011 to 2015.

2 Research framework and hypotheses development

2.1 Research framework

The logical framework of the study is depicted in Figure 1. Admittedly, the current study contends that the correlation among R&D investment and patent application is conditional upon ownership and international expansion. Consequently, the variables need to be jointly analyzed. We engage patent application as output indicators for R&D investment, and comprehensively explores the combined effect of ownership and internationalization on the R&D investment-patent application links to established the clear correlation mechanism among them. The next section discusses the model and the related hypotheses.

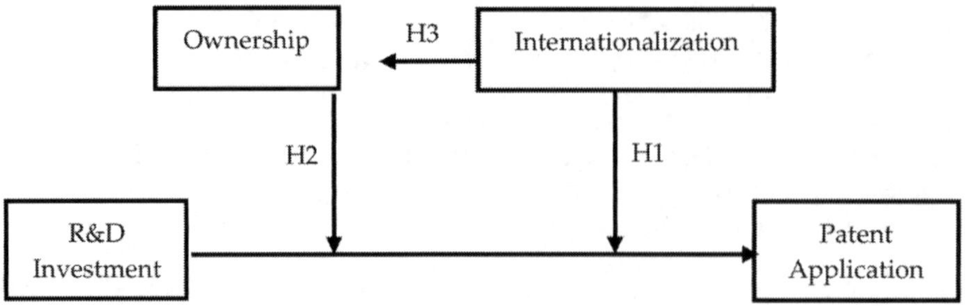

Fig. 1. Research Framework.

2.2 The impact of internationalization on the association among R&D investment and patent application

Internationalization is an essential strategy for emerging market enterprises to achieve technological catch-up and enhance the R&D investment output of enterprises [12]. Unlike developed-economy multinational enterprises, Chinese multinational enterprises need to obtain external heterogeneous resources, advanced technologies, and knowledge for the application of patent innovation in overseas markets. Concurrently, enterprises can organize and learn in the process of internationalization to rapidly improve their innovation abilities to gain market shares as well as a competitive advantage. The reflection of the positive impact among the three variables (i.e., internationalization, R&D investment, and corporate patent applications) is in the following aspects:

First, international expansion can aid enterprises in obtaining external heterogeneous innovation resources, which can provide direction for technology [13], thereby promoting the efficiency of R&D investment. For instance, information resources, R&D personnel, low-priced material, and R&D alliances. As a latecomer entrant to the global economy market arena, emerging economies have very scarce resources in both technology and market [14]. Admittedly, in terms of industrial and technological infrastructure development, emerging economies are far from the global science and technology centers. They also have a disadvantage in terms of accessibility to technical talent resources, innovative information resources, and participation in innovative R&D alliances [15]. These critical restrictions have made it impossible for them to seek resources from the domestic market. Nevertheless, international expansion can help enterprises from emerging economies to obtain low-cost raw materials, labor or tax incentives, and other resources for the reduction of R&D costs [5]. Simultaneously, the international market can encourage enterprises to obtain technical resources that are not available in their home country (e.g., China), but have the potential to help them improve the efficiency of R&D patent output in order to occupy broader market space. Interestingly, when the degree of internationalization of enterprise is low, it is problematic for the enterprise to benefit from R&D investment in a limited market [13]. Conversely, when the degree of internationalization of enterprise is high, the broader market space can effectively share the R&D risk caused by the economic fluctuation of the individual market or the product life cycle, thereby helping enterprises improve the productivity of R&D investment.

Second, international expansion can promote innovative learning on a global scale. Organization learning refers to a process by which an enterprise acquires, absorbs, digests, and applies new knowledge (i.e., external knowledge) about its environment, goals, and processes [16]. The expansion process of internationalization creates an avenue for latecomers to learn from advanced foreign knowledge and experience, cultural intelligence, as well as cross-cultural management to enhance their innovative R&D capabilities. The international market has vast differences (especially with the home market) in terms of resource endowment, technical knowledge, managerial skills, and cultural diversification. The difference has brought a certain degree of negative impact on the enterprise's international expansion strategy. Equally, it has also provided a variety of learning objects for international enterprises, through communication and exchange, learning and imitation, and competition and cooperation with more regions or countries. We believe that it can enhance local enterprises' in-depth understanding of advanced overseas enterprises. Enterprises learning in diversified international environment gain diverse knowledge and innovation experiences to promote their rapid identification of innovation opportunities. Xie, Du, Boadu, and Shi [14] assert that apart from the provision of innovative resources and learning opportunities for enterprises, international expansion can help enterprises to build capability for opportunity identification and opportunity utilization. They also emphasize that enterprises in emerging economies not only learn technology and innovation experience from advanced enterprises in the process of internationalization, but they (enterprises) as well enhanced their knowledge creation performance through reserve knowledge transfer [6], which, in turn, promotes the unit output of enterprises' R&D investment. Hence, we argue that enterprises in emerging economies have benefited from accessing external heterogeneous resources and learning advanced, diversified innovation knowledge and experience through international expansion, thereby promoting unit output of R&D investment. Formally, we propose the following hypothesis: Hypothesis 1 (H1). Internationalization positively moderates the effects of R&D investment and patent application.

2.3 The influence of ownership on the association among R&D investment and patent applications

Prior studies on SOEs and POEs have revealed significant differences in their innovation activities [4]. These two enterprises are associated with different types of ownership; the government / the country owned and controlled the former, whereas an individual or partner owned the latter. From their innovation activities, it is clear that the effects of R&D investments are significantly dissimilar. The impact of corporate ownership in emerging economies on R&D investment output is mainly due to the following aspects:

First, drawing on an agency theory, the separation of ownership and management can create an agency problem. Compared with private enterprises, the agency problem in SOEs is more prominent. In principle, the ownership of SOEs belongs to the entire citizenry of a country. Moreover, there is no real agent and trustee relationship. If the state's supervision mechanism is not perfect, it may be challenging to ensure that the agent will not seek personal gain because of his position [17]. Also, the promotion of senior executives in the SOEs is closely linked to the business performance of the enterprise as well as the political goals of the government. In the pursuance of innovative patents with significant investment, long cycle, and high risk, senior executives in the state-owned enterprises intend to circumvent for hunting such a market-oriented and efficiency-based innovative R&D activity. However, they prefer to implement such opportunities when available in their enterprises. These administrative duties can result in inefficient use of R&D investment [1].

Second, it should be noted that in the state-owned enterprises, political identity is vital for the "convenience conditions" needed to innovate [2]. However, a lot of factors (such as market structure, product structure, technology strategy, technology platform, among others) in non-political identity also serve as a critical element for the realization of state-owned enterprise innovation [18]. Su [19] asserts that the pressure on the executives of state-owned enterprises mainly stems from the market, but the competent authorities. Indeed, it reduces the degree of concern of managers for non-political status, and thus forms a disadvantage of ownership in the innovation efficiency of SOEs. Moreover, state-owned assets holding enterprises usually have internal governance issues such as insufficient resources, managerial corruption, and governmental intervention regarding their administration. These factors have led to inefficient resource allocation. For instance, in state-owned enterprises, the low efficiency of resource allocation makes R&D investment intensity low in terms of innovation performance. Albeit, state-owned assets holding enterprises can support enterprises to obtain more external resources but will hurt innovation efficiency. Formally, we propose the following hypothesis: Hypothesis 2 (H2). State-owned ownership negatively moderates the correlation between R&D investment and patent application.

2.4 The effect of internationalization on the disadvantage of ownership

The management of R&D differs among different ownership types of enterprises. Simultaneously, these differences are affected by the degree of internationalization. We argue that the international expansion of enterprises in emerging economies is conducive to avert the disadvantages of ownership and accelerate the technological catch-up process of state-owned enterprises, thereby increasing the unit output of R&D investment. We elucidate the two main aspects in detail below:

First, from the perspective of internal operations of enterprises, with the deepening of internationalization, state-owned enterprises have realized management modernization and global talent management through international expansion. The process can eventually enable enterprises to make up to some extent the corruption, inaction, and long-term investment commitment of managers in the R&D activities of state-owned enterprises. Thus, promoting the efficiency of R&D input and patent output. As the degree of internationalization increases, companies face fierce competitive pressures leading to higher uncertainty, which eventually lands them to the entrusted agency problems. Therefore, it is essential to improve the incentive mechanism through the adoption of quality management practices and the implementation of effective ways of utilizing resources to ensure that companies enjoy a competitive edge in the global market arena.

Second, from the perspective of external resource utilization, enterprises can obtain vital resources through overseas subsidiaries [20], and transfer these resources to the home country enterprises, to some extent. Such high resources have made up for the problem of the insufficient innovation capacity of state-owned enterprises.

Indeed, emerging market enterprises can enter advanced overseas markets in the process of internationalization, and learn technology, management, operations, and other knowledge, and eventually transfer back to the home country enterprises, which, is conducive to alignment or make up for the weak innovation system in the home country [21]. Precisely, under the conditions of the same scale of the enterprise, the higher the proportion of overseas sales to total sales, the more patent application the enterprise may obtain [22]. Therefore, enterprises actively formulate and implement the strategy of going global, which is conducive to improving the unit output of state-owned enterprises' R&D investment.

Consequently, we argue that state-owned enterprises in emerging economies can be managed to acquire various resources and capabilities essential for innovation through international expansion, thereby diminishing the ownership disadvantages caused by state-owned enterprises. Formally, we propose the following hypothesis: Hypothesis 3 (H3). The negative effects of state ownership on R&D investment and patent application will be weakened through internationalization.

3 Method

3.1 Data and sample

Our study focused on the empirical analysis of secondary data. We selected manufacturing industries from ICT and automobiles as our research objects among all A-share listings on the Shanghai and Shenzhen Stock Exchange for the period 2011- 2015. We collected data on various variables from the CSMAR, WIND database (which are a popular and authoritative database in China in terms of information), and an annual report of listed companies (mainly used to supplement missing data). We obtained 1450 samples from 319 companies. Then, following the rules, we eliminated companies with missing data. As a result, we recorded 1193 valid samples. In all, out of the 319 companies, 242 companies were from the ICT industry, and 77 companies were from the automobile manufacturing industry.

3.2 Measures

3.2.1 Dependent variable

The study adopted a patent application as a dependent variable. To measure the variable, we collected data from the CSMAR database, spanning from the period 2011 to 2015.

3.2.2 Independent variable

The study adopted R&D investment as an independent variable. We used data obtained from the WIND database to measure the variable. Also, due to the changes in a certain period of R&D investment to the patent application, all independent variables are lagged one year to avoid reverse-causality.

3.2.3 Moderating variables

In this study, we have two moderating variables in our statistical model. The variables are internationalization and ownership, respectively. First, we used the ratio of overseas operating income to total operating income as a measure of the DOI. We obtained the data from the WIND. Second, concerning existing studies [23], we consider two dummy variables zero and one as a measure to the ownership characteristics. The value of the state-owned enterprise is 1, and vice versa is 0. We obtained the data from the CSMAR database.

3.2.4 Control variables

We used enterprises' size, age, net profit, the separation of ownership and management, and the industry types as control variables. Based on existing studies, these variables have a significant influence on the internationalization process and technological innovation efficiency [24-26]. The study obtained the information for age and net profit of the companies from the WIND database, whereas, CSMAR database provided data to measure for the scale of the separation of ownership and management, as well as industry types. Besides, the study used the number of workforces to measure the enterprises' size. In summary, the variables of the study are depicted in Table 1.

Table 1. Variables and measurements.

Variable		**Measurement**	**Data Sources**
Dependent variable	Patent application	-	CSMAR
Independent variable	R&D investment	Last year's R&D investment	Wind
Moderator variable	Internationalization Degree	Overseas revenue as a percentage of total revenue	Wind
	Ownership	State-owned enterprises are 1; private enterprises are 0	CSMAR
Control variable	Enterprise size	Number of workforces	Wind
	Age	Year of the establishment - 2015	Wind
	Separation	-	CSMAR
	Industry	ICT industry value 1, automotive manufacturing value 0	CSMAR
	Net profit	-	Wind

4 Empirical results

4.1 Mean, standard error, pearson intercorrelations matrix

Tables 2 reports the mean, S.E, and Pearson intercorrelation matrix of the constructs. The results from Table 2 demonstrate a positive association among R&D investment and patent application ($\beta = 0.589$, $P < 0.01$). It also reveals that enterprise size ($\beta = 0.738$, $P < 0.01$) and net profit ($\beta = 0.663$, $P < 0.01$) have a significant correlation with R&D investment. Besides, we conducted OLS regression on all the models involved in the study to avoid the collinearity problem. Our results show that the VIF values in all models are less than 10, so it can be inferred that there is no obvious multicollinearity in each model of the study

Table 2. Mean, standard error, and correlational relationship among constructs (Cov).

Variables	Mean	S.E	1	2	3	4	5	6	7	8	9
1.Patent application	40.31	224.69	1								
2. R&D investment (million)	190	596	.589 **	1							
3. Internationalization Degree	0.23	0.25	.058 *	-.001	1						
4. Ownership	0.32	0.46	.088 **	.181 **	-.136 **	1					
5. Size (thousand)	6.25	16.07	.344 **	.738 **	.018	.145 **	1				
6. Age	16.72	4.80	.070 *	.067 *	.030	.331 **	.103 **	1			
7. Separation rate	4.40	7.34	.065 *	.041	-.043	.061 *	.016	.094 **	1		
8. Industry	0.76	0.42	.004	-.170 **	.224 **	-.134 **	-.184 **	-.038	-.172 **	1	
9. Net profit (million)	342.76	1791.02	.081 **	.663 **	-.071 *	.141 **	.530 **	.019	.010	-.194 **	1

P<0.05 *, P<0.01 * * (Two-tailed), N=119

Table 3. Regression results and robustness test

Variable/model	Fixed effect model (M1—M4)				2SLS (M2a—M4a)		
	M 1	M 2	M 3	M 4	M2a	M3a	M4a
Size	0.424(0.032)***	-0.143(0.03)***	-0.004(0.028)	-0.177(0.03)***	-0.576(0.04)***	-0.037(0.19)	-0.592(0.039)***
Age	0.024(0.027)	0.02(0.02)	0.04(0.018)*	0.023(0.021)	0.022(0.024)	0.042(0.02)*	0.038(0.025)
Separation rate	0.068(0.028)*	0.035(0.02)	0.019(0.018)	0.026(0.02)	0.014(0.025)	0.021(0.018)	0.005(0.024)
Industry	0.16(0.066)*	0.137(0.049)**	0.055(0.045)	0.131(0.048)**	0.128(0.059)*	0.064(0.055)	0.112(0.058)*
Net profit	-0.132(0.032)***	-0.537(0.027)***	-0.087(0.037)*	-0.413(0.035)***	-0.841(0.035)***	-0.134(0.285)	-0.678(0.044)***
Internationalization Degree			0.052(0.019)**			0.049(0.028)*	
Ownership				-0.007(0.021)			-0.047(0.026)
R&D investment		1.062(0.033)***	0.632(0.04)***	1.069(0.033)***	1.863(0.052)***	0.701(0.423)***	1.831(0.05)***
R&D investment × Internationalization Degree			0.461(0.029)***			0.428(0.200)***	
R&D investment × Ownership				-0.132(0.025)***			-0.154(0.03)***
Cons.	-0.122(0.057)*	-0.104(0.042)*	-0.041(0.039)	-0.076(0.042)	-0.098(0.051)	-0.048(0.046)	-0.058(0.051)
R2 (within)	0.148	0.538	0.621	0.548	0.305 (Centered)	0.618 (Centered)	0.339(Centered)
F	39.080***	229.040***	241.970***	179.070***	249.63***	209.74***	200.96***
Anderson canon. corr. LM statistic					719.575***	392.352***	760.415***
Cragg-Donald Wald F statistic					900.560	289.861	1039.762
Sargan statistic					0.39	0.47	0.63

P<0.05 *, P<0.01 * *, P<0.001 * * *, N=1193, (Two-tailed), Coef. (S.E)

4.2 Hypotheses testing

In this study, we used unbalanced panel data, and mixed OLS regression does not eliminate the interference because the year change has on the regression results. Especially when the same enterprise incorporates repeated observation data across years, it is advisable to use panel data processing methods [4]. At the same time, two categories of models, namely, the fixed-effects model and random-effects model, respectively, in the model selection. Our Hausman test results from M1-M4 show that P-value is > 0.05; therefore, we rejected the original hypothesis. This result demonstrates that the unobserved individual effects are related to the individual effects observed in the model. Thus, the fixed-effect model is selected. Tables 3 and 4 show the results of regression analysis (i.e., the fixed-effects regression analysis). First, Model 1 of Table 3 captures all control variables of the study (F=39.080, P(F)<0.001, R2 =0.148).

Second, the results in M2 show that the enterprise's R&D investment has a statistically significant and positive impact on patent application (β =1.062, P<0.001), which is consistent with the results of most previous studies.

Third, from Table 3, M3, our findings establish that the degree of internationalization of enterprises has a positive moderating effect on the correlation among R&D investment and patent applications (β =0.461, P<0.001). The results indicate that with the intensification of international expansion, the same unit of R&D investment of enterprises can bring more patent applications. Thus, hypothesis 1 is verified.

Third, the interaction term between research and development investment and state-owned ownership has a negative influence on patent applications (β = -0.132, P < 0.001) (see Table 3, M4). Compared with private enterprises, the results indicate that the R&D investment output efficiency of SOEs is low. Thus, hypothesis 2 is verified. To further explore the far-reaching influence of DOI and ownership on the correlation among R&D investment and output, this study further explored in groups.

In Table 4, the study divided the sample companies into two groups according to the degree of internationalization, and also separated the groups into SOEs and POEs. Then, the study formed four sets of data, namely: low internationalization level & SOEs (model M5); low internationalization level & POEs (model M6); high internationalization level & SOEs (model M7); High internationalization level & POEs (model M8). Table 4 reports the impact factors of R&D investment on patent applications in each group. Our results concerning low internationalization show a big gap between SOEs and POEs (see, Table 4). First, the positive effect of R&D investment of SOEs on a patent application (β =0.423, P<0.001, in model M5), is much smaller than that of POEs (β =0.858; P<0.001; in model M6). Second, the positive effect of R&D investment of SOEs to a patent application (β =1.109, P<0.001, in model M7) is more significant and is close to the index of POEs (β =1.169, P<0.001, in model M8). The results demonstrate that spiraling of internationalization can pay-off ownership disadvantage experienced by Chinese enterprises. Thus, hypothesis 3 is verified.

Table 4. Regression result by group.

Variable/model	Lower internationalization M5-M6 (596 firms)		Higher internationalization M7-M8(597 firms)	
	M5 state-owned enterprises	M6 private enterprises	M7 state-owned enterprises	M8 private enterprises
Size	0.117(0.040) **	0.521(0.123) ***	-0.207(0.213)	0.084(0.066)
Age	-0.045(0.033)	-0.342(0.103)	0.199(0.1) *	0.022(0.028)
Separation rate	0.029(0.026)	0.215(0.087)	-0.031(0.118)	0.028(0.028)
Industry	0.085(0.067)	0.699(0.214) *	0.592(0.363)	-0.026(0.07)
Net profit	-0.290(0.044) ***	-0.509(0.134) ***	0.069(0.262)	-0.674(0.123) ***
R&D investment	0.423(0.064) ***	0.858(0.198) ***	1.109(0.214) ***	1.169(0.037) ***
Cons.	-0.048(0.050)	1.864(0.160)	-0.618(0.342)	0.061(0.063)
R2	0.249	0.315	0.308	0.208
F	11.42***	14.53***	10.91 ***	76.69***
N	218	378	158	439

P<0.05 *, P<0.01 * *, P<0.001 * * * (Two-tailed), Coef. (S.E)

4.3 Model robustness check

From the extant literature, scholars attest that endogeneity is a problem that cannot be wholly solved, especially in a panel data analysis for empirical studies. Indeed, endogenous problems in the field of strategic management have received more attention in the literature. Hence, we employed a two-stage least square (2SLS) method to assess the endogeneity of the study. Some authors have pointed out that the 2SLS (two-stage least squares) method is an effective way to correct the endogeneity in the panel data model [27]. The current study considers that the company's research and development capabilities and government subsidies can easily interfere with the correlation among research and development investment and patent application. We utilized the factors as instrumental variables into the 2SLS model for regression.

Table 3 presents the regression results for 2SLS (i.e., values of Model 2a-Model 4a) and the fixed effect model (i.e., values of Model 1-Model 4). From the table, the results for the 2SLS analysis are consistent with the corresponding fixed-effect model results (i.e., from Model 2-Model 4).

At the same time, the study follows [4] and the Stata tutorial to test the validity of the instrumental variables used in the 2SLS model. These include the under-identification test, weak identification test, and over-identification test. First, our findings from Table 3 show that the Anderson-LM statistic of each model of 2SLS is significant (P<0.001), rejecting the null hypothesis of under-identification. Secondly, the Cragg-Donald Wald F statistic is greater than 10 (i.e., the instrumental variable and the endogenous variable have a strong correlation, indicating that the study is suitable for weak identification tests). Thirdly, the Sargan statistic report is not significant (P>0.05), suggesting that there is no over-identification problem. Furthermore, the regression results of the above 2SLS model are consistent with the initial results obtained by the fixed effect model. These findings provide evidence that the conclusion of this study is robust.

5 Discussion and conclusion

In today's ever-changing dynamic environment, innovation has become a lifeblood for firms to accomplish their strategic vision and objectives. Admittedly, firms that posse innovation prowess performs tremendously in the market environment. Interestingly, research and development cannot be ignored in the firm's patent applications. Many firms across the globe spend huge resources on research and development activities to enhance their patent application in order to perform better in the international market arena. Surprisingly, in the extant studies, few scholarships have been executed to examine the stimulating role of research and development investment on the patent application of enterprises from emerging economies. Explicitly, the facilitating influence of DOI and ownership on the linkage between research and development investment and patent application. The current scholarship constructed a logical model to better comprehend the circumstance under which research and development investment influences the patent application of enterprises in China. The findings emanating from the current study have important theoretical implications. First of all, this study explores the influence of the degree of internationalization and ownership on the corporate R&D investment-patent application links. The extant research has paid much attention to the impact of ownership [2-4] and international expansion [5, 6] on innovation performance, but has not reached a consensus conclusion. The main reason may be that ownership and international expansion are considered separately in different models, and the interaction between them is less considered. Based on the existing literature, this study carries out a comprehensive model test, which analyzes the influence mechanism of internationalization degree and ownership on the R&D investment-patent application link.

Second, the current study reveals that internationalization can reduce the negative effect of SOE on the R&D investment-patent application relationship. The unique behavior of Chinese companies in their international expansion is primarily due to their national ownership [7]. Scholars have paid attention to the influence of state ownership on international expansion activities of enterprises [8, 9] and innovation efficiency [10, 11]; however, few studies have focused on the changes in ownership disadvantages as companies continue to expand internationally. Hence, the study explores the changes in the association between state ownership and R&D input and patent output at different levels of internationalization. Our findings demonstrate that the international expansion of enterprises can reduce the negative influence of state ownership on the efficiency of R&D investment. Admittedly, the results indicate that appropriate international expansion of SOEs can compensate for the disadvantages of ownership. Besides, the study discovered and explained an exciting phenomenon and bring forth necessary enlightenment to enterprises in the emerging economies (i.e., represented by China) to improve resource utilization efficiency and achieve technical ability to catch up.

Our empirical results have crucial implications for practitioners and policymakers. First, executives should remain focus and apply all research outputs to address fluctuating market demands speedily. Secondly, policymakers should enact laws and regulations to set a threshold (i.e., lower and upper support limit) on research and development activities and encourage SOEs to adhere to it in their overall goals. Such laws and regulations can lead firms to develop robust innovation performance.

6 Limitation

This paper explores the influence of ownership and internationalization on the R&D investment-patent application link in emerging market enterprises. We adopted the normative method for the study, but the following shortcomings still exist: First, in the sample selection, we considered only two industries, namely; ICT and automobile manufacturing, as samples.

Our overall sample size is small, and the expansion of the industries may lead to further discoveries. Therefore, future research can consider a broader sample size to test the research model. Second, our study only explores the influence of the degree of internationalization and state ownership on the R&D investment-patent application link. The perspective is relatively simple. There is no further discussion on the interference factors such as the monopoly of SOEs and the political purpose of the SOEs in the process of internationalization. We encourage future researchers to consider interference factors. Third, the study focused on China. A future study can examine the generalizability of our findings to other emerging economies.

Acknowledgments

This research was financially supported by the High level talents Research Foundation Project (No. 950319146); and National Natural Science Fund of China (No.71672021) .

References

1. K. Z. Zhou, G. Y. Gao, H. Zhao, State Ownership and Firm Innovation in China: An Integrated View of Institutional and Efficiency Logics. Administrative Science Quarterly, **62**, 2 (2017)
2. V. Z. Chen, J. Li, D. M. Shapiro, X. Zhang, Ownership structure and innovation: An emerging market perspective. Asia Pacific Journal of Management, **31**, 1, 1-24 (2014)
3. P. Boeing, E. Mueller, P. Sandner, China's R&D explosion—Analyzing productivity effects across ownership types and over time. Research policy, **45**, 1, 159-176 (2016)
4. Y. Yang, J. Wei, L. Luo, Who is using government subsidies for innovation? - The joint adjustment effect of ownership and factor market distortions. Management world, **1**,1, 75-86 (2015)
5. B. Hai, n. Nie, Internationalization, Innovation and Corporate Performance: An Empirical Study Based on Hubei Province. Scientific research management, **33**, (4), 1-9 (2012)
6. L.H. Du, Creative asset seeking, subsidiary influence and reverse knowledge transfer. Scientific research management, **3**, 39, 85-96 (2018)
7. R. Ramamurti, J. Hillemann, What is "Chinese" about Chinese multinationals? Journal of International Business Studies,**49**, 1, 34-48 (2018)
8. M. W. Peng, S. L. Sun, B. Pinkham, H. Chen, The institution-based view as a third leg for a strategy tripod. Academy of Management Perspectives, **23**, 3, 63-81 (2009)
9. Y. Luo, R. L. Tung, A general theory of springboard MNEs. Journal of International Business Studies, **49**, 2, 129-152 (2018)
10. K. Z. Zhou, G. Y. Gao, H. Zhao, State ownership and firm innovation in China: An integrated view of institutional and efficiency logics. Administrative Science Quarterly, **62**, 2, 375-404 (2017)
11. X. Cao, D. J. Cumming, S. Zhou, State ownership and corporate innovation efficiency. Available at SSRN 2868036 (2018).
12. B. Hagen, S. Denicolai, A. Zucchella, International entrepreneurship at the crossroads between innovation and internationalization. Journal of International Entrepreneurship, **12**, 2, 111-114 (2014)

13. M. A. Hitt, R. E. Hoskisson, H. Kim, International diversification: Effects on innovation and firm performance in product-diversified firms. Academy of Management journal, **40,** 4, 767-798 (1997)
14. Y. Xie, Y.-F. Du, F. Boadu, X.-Y. Shi, Executives' Assessments of Evolutionary and Leapfrog Modes: An Ambidexterity Explanation Logic. Sustainability,**10**,8,2893(2018)
15. J. A. Mathews, Dragon multinationals: New players in 21 st century globalization. Asia Pacific journal of management, **23**, 1, 5-27 (2006)
16. W. M. Cohen, D. A. Levinthal, Absorptive capacity: A new perspective on learning and innovation. Administrative science quarterly, **12**,1,1128-152 (1990)
17. A. I. Khwaja, A. Mian, Do Lenders Favor Politically Connected Firms? Rent Provision in an Emerging Financial Market. Quarterly Journal of Economics, **120**, 4, 1371-1411 (2005)
18. J. Su, Q. Lin, Independent innovation of state-owned enterprises: What are the situational factors besides political identity? Management review, **28**, 3, 230-240 (2016)
19. X. Su, Managers' Work Stressors and Their Impacts—Comparison between State-owned Enterprises and Private Enterprises. Management world, **2** (8), 105-113 (2005)
20. Y. Luo, R. L. Tung, International expansion of emerging market enterprises: A springboard perspective. Journal of international business studies, **38**, 4, 481-498 (2007)
21. A. Madhok, M. Keyhani, Acquisitions as entrepreneurship: Asymmetries, opportunities, and the internationalization of multinationals from emerging economies. Global Strategy Journal, **2**, 1, 26-40 (2012)
22. P. Zeng, T. Deng, The relationship between the degree of internationalization of enterprises and technological innovation: a perspective of learning. International trade issues, **3**,10, 59-67 (2012)
23. M. Kotabe, S. S. Srinivasan, P. S. Aulakh, Multinationality and firm performance: The moderating role of R&D and marketing capabilities. Journal of international business studies, **33**, 1, 79-97 (2002)
24. E. Huergo, J. Jaumandreu, Firms' age, process innovation and productivity growth. International Journal of Industrial Organization, **22**, 4, 541-559 (2004)
25. F. Damanpour, Organizational size and innovation. Organization studies, **13**, 3, 375-402 (1992)
26. R. Ortega-Argiles, R. Moreno, J. S. Caralt, Ownership structure and innovation: is there a real link? The Annals of Regional Science, **39**, 4, 637-662 (2005)
27. Y. Wang, H. Li, Endogenous problems and correction methods in management research. Management quarterly, **2**, 3, 20-47 (2017)

The Embedded Strategy of Enterprise Innovation Network Based on Knowledge Base

Chen Guang Li*, *Zhenjun* Qiu and *Guihuang* Jiang

Department of Administration Management, School of Economics & Management, North China, University of Technology, Beijing, China

Abstract. Constructing the innovation network is an important means of breakthrough in the traditional manufacturers for the emerging e-commerce platform, based on two different levels of innovation network node, content and structure, YEATION and YOUPIN were chosen to analyse and research the relationship and between knowledge base and the embedding strategy of electric business platform enterprises to create innovation network. To put forward the knowledge base width wider, insufficient depth of the enterprise should adopt structural strategy give priority to relational embeddedness, on the other hand, enterprises of which the depth of knowledge base is deep and width of insufficient shall adopt strategy give priority to the embedding of structure. At the same time, it is proposed that platform enterprises should not adopt a single embeddedness strategy, but should adopt auxiliary embeddedness strategy in a timely manner.
Keywords: Innovation network, Knowledge base, Embeddedness.

1 Introduction

As Chesbrough [1] suggested the way products innovated and promoted have been undergoing a reform during the past decades, manufacturers pursuing a expansion toward higher value-add link. The traditional model of enterprises completing development, manufacturing, marketing, supply and service independently becoming more and more difficult, especially for latecomers in a network circumstance. Prior literature has developed multifarious conceptual models to illustrate the pathway, for example, the "flying geese" paradigm [2], the "imitation to innovation" transformation [3,4], among which the track from original equipment manufacturing (OEM) to original design manufacturing (ODM) and original brand manufacturing (OBM) has been one of the most interested among scholars [5-7]. To progress to OBM, manufacturers need to possess the skill and resource of marketing [8,9].

Under innovation network background, internet companies adopt quality e-commerce, which can be seen as integrated vertical e-commerce, and achieve building industrial barriers, competitive brands, and embedding in innovation network by cultivating ODMs into OBMs. In the last decade, many quality e-commerce emerged and achieved financial returns, for instance, YANXUAN, YOUPIN, XINXUAN, JINGZAO, and so on. This paper study how

* Corresponding author: chilqzj@163.com

quality e-commerce integrate and exploit knowledge base of its OBMs in embedding innovation network

This article selected YANXUAN and YOUPIN as typical quality e-commerce, aiming at contributing to the debate on the relationships between innovation, industrial dynamics and industry evolution, by analyzing the links between the knowledge-base and embeddedness.

2 Research background

The innovation process was defined as a search process of inventions generated from recombination of existing knowledge elements novelty [10]. In this search process, the knowledge couplings a firm can reach, which is merely a small subset of the knowledge arena, is defined as knowledge base [11]. The expansion of knowledge can be labeled as two patterns, 'widening' and 'deepening'. The former is associated with knowledge searching distally across disciplines or fields, while the latter is characterized by accumulation strategies based on the existing vertical premises. Former study reveal that depth and breadth of knowledge base are important for firm performance and innovation [12,13]. More specifically, empirical Study revealed that firms' knowledge acquisition and subsequent innovative process were significantly influenced by their knowledge base characteristic [14]. Knowledge base can be either premises and constraint of particular innovation process. Zhou and Li proposed that firms with a broad knowledge base tend to achieve radical innovation through internal knowledge sharing rather than market knowledge acquisition, while firms with great knowledge base depth are more capable of developing in opposite way [15]. Carnabuci and Operti [16] indicated that knowledge of great breadth hinders firm's recombinant reuse and improve it ability to innovate through recombinant creation, knowledge base depth influencing on the opposite way. Prior Study provides evidence regarding the relationship between change in knowledgebase characteristics and industry structure in the pharmaceutical industry [17].

Twenty years after Granovetter introduce embeddedness into social network, it has developed two main perspectives: structural and relational. Structural embeddedness shows the impact that a firm's relative position in the network has on performance. Structural Perspective emphasizes properties of ego-network (focal firm) or whole network such as structural hole [18], density [19], centrality [20], size [21]. Relationship embeddedness reflects the degree of trust and information exchange between two innovators in a network, usually focus on the strong ties and weak ties within network [22].

Study of knowledge base and properties of embeddedness about what influence they have on performance show a tendency toward contingency. Kang and An [23] proposed that structural embedding is beneficial for firms in quantity of resources, while relational embedding benefits firms obtaining better resources. Characteristics of network have complex influence on performance [24,25]. This study attempts to investigate the embedding path selection of firms under different knowledge bases through comparative case studies.

3 Case study

3.1 Sample and data collection

This paper involves the selection of strategies for external network embedding based on the internal knowledge base of the enterprise, selecting comparative case study as study method as which is a question of know-how. Of all the quality e-commerces in China, we chose typical and relatively mature ones as case firms: YANXUAN and YOUPIN, who both stablished recognizable brand styles while of distinctly different knowledge base. The basic information is as follows.

The data sources for this paper are mainly from e-commerce platform websites and disclosures, reports and interviews collected from the Internet.

3.2 YANXUAN

This paper involves the selection of strategies for external network embedding based on the internal knowledge base of the enterprise, selecting comparative case study as study method as which is a question of know-how. Of all the quality e-commerces in China, we chose typical and relatively mature ones as case firms: YANXUAN and YOUPIN, who both stablished recognizable brand styles while of distinctly different knowledge base. The basic information is as follows.

YANXUAN as for the upstream of the value chain, YANXUAN adopts a mechanism called ODM buyer, to eliminate intermediate agents and to exert control over raw materials, production, QA and post-sale. The buyers select and contract excellent ODMs, especially manufacturers of top-tier international brands including that of Muji, Uniqlo, Samsonite, Hermes, Shiseido, CK, Nine West, Armani, Ralph Luaren, Chloe, Timberland, Fendi, Stoneline, Coach, HUGO BOSS, Under Armour.

In the midstream, YANXUAN utilizes flexible supply chain to cope with expanding categories. Within six mouthes from the launch of YANXUAN, its products include nine primary categories, 52 subcategories and 592 SKUs (Stock Keeping Unit), with is already wide coverage compared to general start-up e-commerce. One year later, SKUs climbed to over 10000 in a surprising speed. Unlike other large e-commerce' "space for time" solution of dense warehousing network, YANXUAN's flexible supply chain subdivides production plans into batches of smaller quantity. The launch of new products firstly commissioned a few batches, additional orders were adjusted sensitively according to the front-line sales volume. In an emerging of hot cake, it can also leverage control over supplier to fully utilize the production line and meet market demand timely.

From the sales side of the platform, YANXUAN cultivates the user group with the brand concept of "good life, not so expensive". As Walsh and Mitchell [26] argued, with the introduction of high quality labeled private merchandise, the confusion proneness associated with quality and emotions have significant impact on marketplace. By combining the user group portraits of NetEase Open Class and NetEase Cloud Music, YANXUAN targets its core user group at people aged 20-35 years old who seek to improve their quality of life but have limited purchasing power and define the brand style as a mid-range brand rich in artistic temperament, pursuing sophistication and advocating lifestyle aesthetics.

3.3 YOUPIN

YOUPIN intervenes in the upstream supply chain through investment and incubation. Backed on Xiaomi's mobile phone and IOT (Internet of Things) , which has been the largest smart hardware platform worldwide, YOUPIN incorporates cooperative enterprises into the Xiaomi ecological chain, exerting values and participating in controlling design concepts, product definition, industrial design and other aspects.

YOUPIN differ from other quality e-commerces mainly in three principles of investment and incubation. Firstly, it cultivates ODMs by exporting enterprise management methodology and product development methodology, providing resources like channels, clients and even supply chain and moreover, values. Secondly, it has skin in the game with ecosystem companies through equity participation but not holding in order to maintain the combativity and innovation of startup teams. Finally, manufacturer can compete with products of YOUPIN. As Hoch and Banerji [27] proposed, private label goods should complement, not replace, the manufacturer's brand; that the right choice is for retailers to offer high quality and relatively low prices; and that private labels should try to avoid face-to-face competition with manufacturer-branded goods. Typical example is Huami, being the world's leading wearable maker, differentiating its main product AMAZFIT series from Xiaomi bracelets. Most different from YANXUAN, YOUPIN strives to support manufacturers in establishing their own independent brands, and the eco-friendly Huami, Zmi, Yeelight, Smartmi, Aqara, etc are already at the forefront of their respective industries. At the horizontal platform level, YOUPIN takes the lead in getting manufacturers to support reciprocally. Zmi, after becoming a leader in the mobile power industry, gradually provides battery for other ecological enterprises and help them improve power management solutions, getting stable customers in return.

3.4 Knowledge base

This paper, specific to the study of quality e-commerces, illustrated the knowledge base as primarily the ability to leverage its own brand resources and select manufacturers as well as products. This paper argues that YANXUAN is superior in terms of knowledge base breadth, while YOUPIN advanced in knowledge base depth.

As NetEase, the parent company of YANXUAN, has accumulated experience in grasping user groups, Internet marketing and establishing brand tone through its community-based products such as open classes and music while lacking experience about goods. On the other hand, the ability to select products need to be more proficient in the mastery of the manufacturing field, YOUPIN's knowledge from Xiaomi of the field of technology products has a more advantageous, more competitive than YANXUAN of the household merchandise.

3.5 Embeddedness.

YANXUAN adopted an embedding strategy that was first structural then relational, but did not achieve relational embedding successfully. After scaling back operations, YANXUAN still has far more SKUs on display than its competitors. In spite of strong supply chain management capabilities, the relationship between YANXUAN and ODM manufacturers stays in the contract.

The rapid expansion of the product system makes YANXUAN quickly occupy the central position of the ODM network, but the speed of experience accumulated by the category team can't keep up with the speed of SKU expansion, and can't rely on the relationship established with manufacturers to achieve complementary experience, the control of suppliers is insufficient. In addition, the product itself is neglected and the quality control capability is lost. Limit to contractual relationship, YANXUAN cannot control the manufacturer's R&D activities, and ODM only provides homogeneous products according orders, which to some extent constraints the manufacturer's ability to stimulate innovation.

YANXUAN manages suppliers and a large number of SKUs through Tongzhou, an intelligent cloud collaboration platform for suppliers, which enables sharing of raw material stocking plans, production progress, inventory sharing, standardized quotation process, raw material price reference, as well as supplier optimization and assessment based on the above functions to improve the daily management of suppliers' sales. Tongzhou enables YANXUAN get in touch with ODMs as many as it want. Despite the brand's clear style, there is not too strong a definition of the user group. YANXUAN has opened a product recommendation of styles from Japan, France, Thailand, New Zealand, Vietnam, and Korea, and works with a large number of professional designers and design companies at home and abroad.

YOUPIN, on the other hand, chose a relational to structural embeddedness strategy. About three quarter of the suppliers on YOUPIN come from Xiaomi or its ecosystem companies, and YOUPIN has established close ties with them in many ways, including R&D, patents, sales, QA, inventory, logistics, accounting processing, and IOT ecology etc.

Constrained by the brand tone of the Xiaomi brand, it is difficult to expand consumer base and find added value in terms of emotional value. In the face of surging demand, YOUPIN is incapable of expanding its subcategories with guaranteeing quality.

Though YOUPIN's initial brand position of pursuing cost-effective and IOT ecology makes its consumer groups solid but difficult to expand, forcing the ecological chain companies to try untouched high-end market of YOUPIN itself, forming a virtuous cycle. Successful examples includ Huami's Amazfit and Zmi's Zmi 10 etc.

Through a comparative study of the above cases, this paper suggests that the path of embedding strategy should take into account the internal knowledge base of the platform-type enterprises in this paper, enterprises of which the breadth of the knowledge base is wider and lack in depth should adopt a structured strategy, focusing on relational embedding strategy timely; on the other hand, enterprises of which knowledge base is with great breadth and insufficient depth should adopt a relational strategy, gradually structured embedding strategy, instead of a single, consistent embedding strategy.

4 Conclusion

This study, based on a comparative case study of YANXUAN and YOUPIN, argues that the selection of embedded strategies for SMEs based on the internal knowledge base of the enterprise has some implications for the selection of network strategies for SMEs.

Various paradigms between OEM, ODM and OBM are emerging, such as many new formats like "new retail". The relationship between enterprise knowledge heterogeneity and its linkages also needs to be further discussed. The pursuit of differentiation among competing firms, the study of their development direction can more effectively reflect the dynamic and contingent characteristics of innovation network research than focusing merely on performance.

The case subjects selected for this paper are small, the research model used may be oversimplified, and future research will require a larger sample of data.

Acknowledgements

This research was financially supported by the Humanities and Social Sciences Fund Project of the Ministry of Education of China (No.20YJCZH066), the Beijing Municipal Social Science Fund (No.16GLC041), the National Nature Science Foundation of China for Youth (No.71503011), the Beijing Municipal Young Top-Notch Talent Funding Program (2017000026833ZS02) and the Basic Scientific Research Business Expenses Project of Beijing Municipal Education Commission (No.110052971921072).

References

1. H. Chesbrough. The logic of open innovation: managing intellectual property, California management review, **45**, 33-58 (2003)
2. K. Akamatsu, A historical pattern of economic growth in developing countries, The developing economies, **1**,3-25, (1962)

3. D.Kale, L. Steve Little. From imitation to innovation: the evolution of R&D capabilities and learning processes in the Indian pharmaceutical industry, Technology Analysis & Strategic Management, **19**, 589-609 (2007)
4. F. Wang, K. Chen. Do product imitation and innovation require different patterns of organizational innovation? Evidence from Chinese firms, Journal of Business Research. **106,** 60-74 (2020)
5. M. Hobday. East Asian latecomer firms: learning the technology of electronics, World development, **23,** 1171-1193 (1995)
6. K. Lee. Making a technological catch - up: Barriers and opportunities, Asian Journal of Technology Innovation, **13**, 97-131 (2005).
7. W. Chu. Can Taiwan's second movers upgrade via branding, Research Policy, **38**, 1054-1065 (2009)
8. J. Weerawardena. The role of marketing capability in innovation-based competitive strategy, Journal of Strategic Marketing, **11,** 15-35 (2003)
9. A. Kaleka, A. M. Neil. How marketing capabilities and current performance drive strategic intentions in international markets, Industrial Marketing Management, **78**,108-121 (2019)
10. F. Lee. Recombinant uncertainty in technological search, Management Science, **47,** 117-132 (2001)
11. S. Yayavaram, W. Chen. Changes in firm knowledge couplings and firm innovation performance: the moderating role of technological complexity, Strategic Management Journal, **36,** 377-396, (2015)
12. S. Moorthy, D. E. Polley. Technological knowledge breadth and depth: performance impacts, Journal of Knowledge Management, **14,** 359-377 (2010).
13. S. Awate, M. M. Larsen, R. Mudambi. EMNE catch - up strategies in the wind turbine industry: is there a trade - off between output and innovation capabilities, Global Strategy Journal, **2** 205-223 (2012)
14. Subramanian, Annapoornima M., Wang Bo, and Chai Kah-Hin. "The role of knowledge base homogeneity in learning from strategic alliances." Research policy **47** (1) 158-168 (2018)
15. K. Z. Zhou, C. B. Li. How knowledge affects radical innovation: knowledge base, market knowledge acquisition, and internal knowledge sharing, Strategic Management Journal, **33,** 1090-1102 (2012)
16. G. Carnabuci, E. Operti. Where do firms' recombinant capabilities come from? Intraorganizational networks, knowledge, and firms' ability to innovate through technological recombination, Strategic Management Journal, **34,** 1591-1613, (2013)
17. J. Krafft, Q. Francesco, P. P. Saviotti. Knowledge characteristics and the dynamics of technological alliances in pharmaceuticals: empirical evidence from Europe, US and Japan, Journal of Evolutionary Economics. **24,** 587-622 (2014)
18. T. Lou, T. Jie. Mining structural hole spanners through information diffusion in social networks, Proceedings of the 22nd International Conference on World Wide Web. 2013.
19. M. A. Schilling, C. C. Phelps, Interfirm collaboration networks: the impact of large-scale network structure on firm innovation, Management Science. **53,** 1113-1126 (2007)
20. T. Caner, S. Jing, E. P. John. When a firm's centrality in R&D alliance network is (not) the answer for invention: the interaction of centrality, inward and outward knowledge transfer, Journal of Engineering and Technology Management. **33**, 1113-1126 (2014)

21. N. Lahiri, N. Sriram. Vertical integration, innovation, and alliance portfolio size: implications for firm performance, Strategic Management Journal. **34,** 1042-1064 (2013)
22. M. Tortoriello, R.R. Marco, B. McEvily. Bridging the knowledge gap: the influence of strong ties, network cohesion, and network range on the transfer of knowledge between organizational units, Organization Science. **23**, 1024-1039 (2012).
23. S. Kang, L. An. Network embedding, innovation ability and knowledge power: a perspective based on global value chain, Science of Science and Management of S.&.T. **40,** 1-12 (2019)
24. V. Gilsing, N. Bart. Density and strength of ties in innovation networks: an analysis of multimedia and biotechnology, European Management Review. **2**, 179-197 (2005)
25. J. Zhang. Facilitating exploration alliances in multiple dimensions: the influences of firm technological knowledge breadth, R&D Management. **46**, 159-173 (2016)
26. G. Walsh, V. W. Mitchell. The effect of consumer confusion proneness on word of mouth, trust, and customer satisfaction, European Journal of Marketing, **44,** 838-859 (2010)
27. S. J. Hoch, S. Banerji. When do private labels succeed, MIT Sloan Management Review. **34** (1993)

Empirical Research on the Influence of International Crude Oil price fluctuation on Russian Exchange Rate

Qingyao Li *

School of Finance, Xinjiang University of Finance and Economics, China

Abstract. Oil is a kind of basic energy and has close relationship with the financial markets. Russia is the world crude oil exporter, whose rich energy resources make the country tend to invest in energy industry and make policies for it. This has led to the fact that its national economic structure is too simplified and its economy is heavily dependent on the oil industry. The total oil exports for Russia each year account for up to forty percent of its fiscal revenues. The economic characteristics of Russia and the fluctuation of international crude oil price greatly influence the macroeconomic changes of a country and therefore the exchange rate will make a change too. So this paper, will explore the relationship between the international oil price and Russian exchange rate and the degree of the influence between each other. The paper will select international crude oil price as exogenous variables and take the Russian currency as endogenous variables to establish VAR model. In view of the conclusion, this paper points out the drawbacks of Russia's current economic development, and finally puts forward reasonable and effective suggestions from the aspects of exchange rate system and economic structure.

1 Introduction

Domestic and foreign research on the correlation between crude oil price and exchange rate [1] Junling Zhang and Weixian Wei (2017) used the dynamic GTAP model to simulate the transmission effect of the double impact of international oil price fluctuation and RMB devaluation on the macro economy. Empirical results show that international oil price fluctuations will have a greater impact on national macroeconomic. [2]Yu Wang and Da Wu(2017) selected data of several oil-producing countries as samples to list their exchange rate policies and make comparisons. Under the background of fluctuations in international crude oil prices, the main factors affecting the selection of exchange rate policies of sample oil-producing countries include the foreign exchange reserves of a country and the degree of national economy's dependence on oil.[3]Siqing Zhuo,Bo Wang and Lu Qiao (2017) used the Coupla function model as the framework to study the influence relationship between changes in international crude oil prices and Russia's real exchange rate.The conclusion shows that there is a strong left-right tail dependence between the two variables, which proves

* Corresponding author: 1272679496@qq.com

that the sharp fluctuation of Russian exchange rate in recent years is largely caused by the drastic fluctuation of international oil price.[4]Xi Zhang (2011) used co-integration analysis, concluded that the international crude oil price was a significant financial influence factor on the RMB exchange rate, and would also have an impact on the domestic fuel industry and its related products.[5]Yi Zhang,Ke Wang and Lin Jiang (2014) introduced three variables, namely OECD commercial inventory level, US industrial output coefficient and federal funds rate, and they finally reached a conclusion: the US dollar exchange rates had different degrees of influence on oil price in different periods.

In the existing studies, most of scholars mainly focus on the relationship between the rising international crude oil prices and the economic changes happening several years ago, Among them, the amount of theories is more than quantitative analysis on the relationship between the two factors and the conclusion mostly certifies that international oil prices to a certain extent will lead to the rising production cost for most of processing industry.It is predicted that the unemployment rate will increase and therefore inflation will follow. In that case, their economic growth will be hindered and with the degree of trade globalization increasing, it will make the adverse effect spread among nations. Moreover, most of the studies focus on the impact on the US dollars, major developed countries and the macro economy of RMB while there is little research on Russia, a big oil producer in the developing world. However, the situation of ruble depreciation and inflation caused by the sharp drop in oil prices since 2014 contradicted many verdicts. Therefore, in order to explore the relationship between oil price and exchange rate and the degree of mutual influence, this paper establishes VAR model.

2 Model construction and data sources

2.1 Model construction

Eviews7.2 was used to carry out empirical research on 109 samples of relevant data from 2010 to 2019. Two variables were adopted: explanatory variable, International crude oil price OP, and explained variable, Russian Real effective exchange rate EX. In order to reduce heteroscedasticity-autocorrelation and increase the stability of the sequence, the logarithm of variable data was used to establish the following Var model:

$$lnEX_t = C_1 + \sum_{i=1}^{q} \boldsymbol{\alpha}_i lnEX_{t-i} + \sum_{i=1}^{p} \boldsymbol{\beta}_i lnOP_{t-i} + \cdots + u_{1t} \quad (1)$$

$$lnOP_t = C_2 + \sum_{i=1}^{p} \boldsymbol{\theta}_i lnEX_{t-i} + \sum_{i=1}^{q} \boldsymbol{\gamma}_i lnOP_{t-i} + \cdots + u_{2t} \quad (2)$$

Where, C_1 、C_2 are the constant terms, u_1t 、u_2t are the independent white noise sequences, and α_i 、β_i 、θ_i、γ_i are the coefficient of vectors. These significance represent the correlation between vectors, and i is the lag order.

2.2 Description of variables

Russia belongs to the territory of Europe. The price of domestic Urals crude oil depends on the trading price of British Brent crude oil, so the price of Brent crude oil is used to replace the international crude oil price. The price of Brent crude oil is an independent variable determined by the model.The Russian real effective exchange rate represents the Russian exchange rate. This variable can more accurately measure the purchasing power of a country's currency after excluding the factor of inflation, and the data are easier to search, thus establishing a stable external reference condition for the model.

Monthly data of Brent crude oil price is based on the closing price of Brent Futures Index in Sina Finance and Economics from November 2010 to the end of November 2019. Monthly data on the Russian real effective exchange rate are collected at the end of each month for the same period in Eurostat.

3 Empirical analysis

3.1 ADF test

The purpose of studying time series is to obtain information from historical data. Data stabilization represents time translation invariant to some extent. ADF test determines whether unit root exists in a sequence.The logarithmic transformation of the original data represents the change rate of each variable in the constructed model, which has its research significance. Meanwhile, heteroscedasticity and autocorrelation are reduced to increase the stationarity of the sequence.As shown in Table 1, the sequence ln(OP) is unstable, while ln(EX) is stationary. However, if these variables are taken as first-order differences containing constant terms, two groups of stationary sequences can be obtained.In addition, after the regression, the residual sequence was extracted for the same test, and it was found that the stationary sequence represented the co-integration relationship between variables, namely, the long-term equilibrium relationship.

Table 1. ADF test of international oil price and Russian exchange rate.

Variable	Inspection form	The t-statistics	Prob.*	The results
in (EX)	trend and intercept	3.479530	0.0468	smooth
dln (EX)	intercept	-7.314090	0.0000	smooth
ln (OP)	intercept	1.488626	0.5357	not smooth
dln (OP)	intercept	8.299678	0.0000	smooth
e_i^2	intercept	3.822767	0.0037	smooth

Note: Dln (EX) is the first difference of Ln (EX)

3.2 Granger test

Since these two time series are not integral of the same order, Granger test is conducted on the variables after the simultaneous first-order difference.The test results in Table 2 below show that DLN (EX)- DLN (OP) is true, and DLN (OP)- DLN (EX) is also true.That is, $P<0.05$ rejects the null hypothesis and there is a bidirectional Granger causality, indicating that the rate of change of international oil price and the rate of change of Russian real exchange rate can respectively explain the change of each other.

Table 2. Granger causality test of the impact of the rate of change in international crude oil prices on the rate of change in Russian exchange rate.

The null hypothesis	F-statistic	Prob.
Dln(OP) does not Granger cause Dln(EX)	3.95654	0.0222
Dln(EX) does not Granger cause Dln(OP)	4.36725	0.0152

3.3 The determination of the optimal lag order

According to the test results of the stability of the above data, the first-order difference of two variables ln(EX) and ln(OP) can be studied empirically by using the vector autoregression model. Q criterion and SC criterion select the optimal lag order as the first order, while F statistic criterion and AIC information criterion select the third order as the best. Moreover, because the number of samples is large, the third order lag will not cause serious loss to the data. Therefore, the comprehensive evaluation decides to select the optimal lag order P =3 to build the VAR(3) model.

3.4 Impulse response

When the stationarity of the VAR (3) model was measured, it was found that all points were obviously located in the unit circle, indicating that the VAR model was stable. Based on this premise, the impulse response was established as shown in the figure below to obtain the following conclusions.

In Figure 2, it can be seen from the impulse response of DOP to DEX that DLN (OP) gives a positive impact in the first phase and reaches the maximum impulse response value of 0.008 in the third phase. Then it goes negative in phase 4, reaches its maximum in reverse in phase 5, and then gradually approaches 0.This indicates that after the international oil price shock, the real exchange rate of Russia had a significant positive impact before the fourth period, reflecting a small depreciation, but the positive impact only lasted for four periods.

From the impulse response of DEX to DOP in Figure 3, it can be seen that when ln(EX) gives ln(OP) a positive impact, the impulse response value of the rate of change of international oil price reaches the maximum value of 0.026 in the second phase, and then starts to continue the negative impact, and gradually shrinks back to 0 point after reaching -0.09 in the fifth phase. This means that the positive impact of the rate of change of the Russian exchange rate on the rate of change of the international oil price has a shorter duration than the effect of the oil price on the real exchange rate, but the impact of the rate of change of the oil price has a greater impact on the rate of change of the rouble exchange rate.

Figure 1 and Figure 4 respectively show the impulse response function of the rate of change of exchange rate of Russia and the rate of change of international oil price to itself. It can be seen that in the first phase, the impulse response of Dln(EX) to itself reached the maximum value, which was 0.33, and then gradually moved to -0.09. Finally, the impulse effect disappeared. The rate of change of international oil price has a positive impact on itself in the first stage as high as 0.72, and then fluctuates in a small range around 0 from the fourth stage.

On the whole, the trend of the real exchange rate fluctuation of Russia is a small depreciation followed by a small appreciation. The reason may be that at the beginning, due to the impact of the international oil price, the total export of Russian crude oil declined rapidly while the import elasticity was relatively small. The rising inflation made the exchange rate fluctuation in previous periods intensified and led to depreciation. After the impact, the international demand for oil increased, the export recovered gradually, and the real exchange rate appreciated. Russia's currency for Brent oil prices is slightly greater than the influence of oil prices on the impact of exchange rate. It may be due to the fact that Russia is a energy giant in its oil reserves, production, export and transit and it plays a role in the international petroleum import and export pattern and pricing mechanism, so in the short term, Russia's real exchange rate fluctuations impact results are shown in the process in which the oil prices first rise and then fall.

 https://doi.org/10.1051/shsconf/20219604006

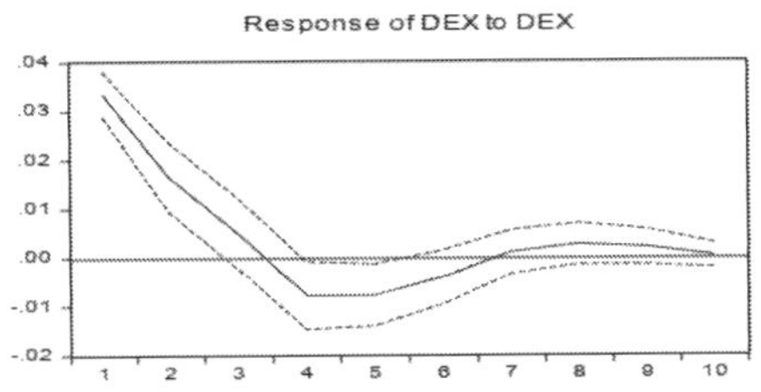

Fig. 1. Response of DEX to DEX.

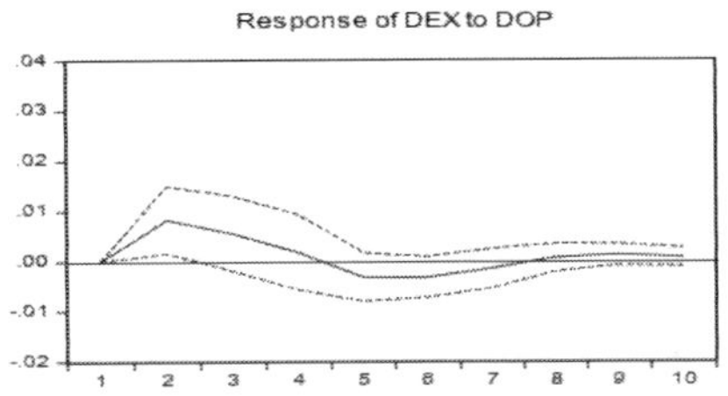

Fig. 2. Response of DEX to DOP.

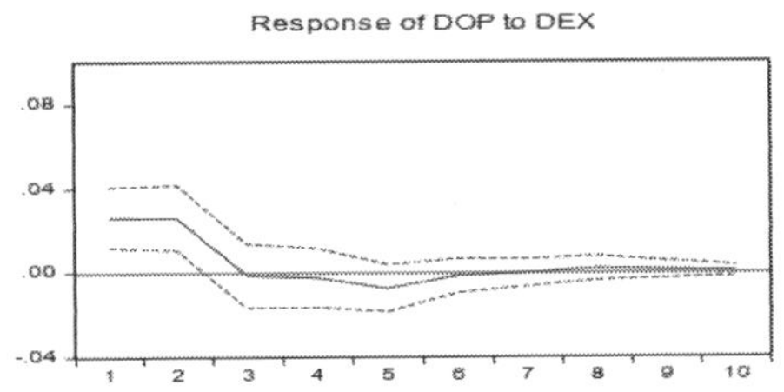

Fig. 3. Response of DOP to DEX.

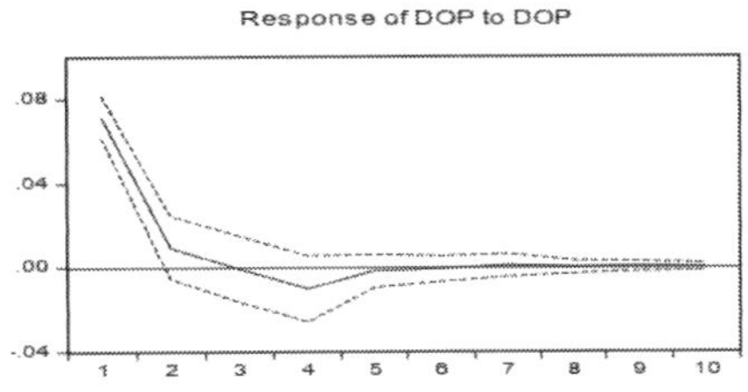

Fig. 4. response of DOP to DOP.

4 Conclusions and suggestions

4.1 Conclusion

Observation of the above empirical analysis results can be summarized as follows: First, the ADF test shows that the first-order difference between the international crude oil price and the Russian exchange rate is stable and there is a long-term equilibrium relationship between the original series. Second, Granger causality test shows that one of the changes in the international crude oil price and the Russian exchange rate can explain part of the changes in the other side. Finally, the impulse response results show that in the short term, the impulse response effect of the international oil price on the real exchange rate of Russia lasts longer, but the rouble exchange rate can have a greater shock on the international oil price.

Russia is a typical energy-dependent exporting country, with oil exports accounting for about 40 percent of its national revenue each year.Therefore, the changing factors in the international economic environment are more likely to make the rouble exchange rate fluctuate. When the international oil price is high, the rouble exchange rate will be stable, which may lead to a slight appreciation. When international events lead to lower oil prices, the depression of the price will often lead to drastic fluctuations and sharp drops in the rouble exchange rate.It can also be observed from the annual chart that the rise and fall of crude oil export prices and rouble exchange rate fluctuations show a more obvious consistency.The empirical evidence shows that the crude oil price has a strong correlation with the exchange rate of Russia, which also reflects the disadvantages of Russia's economic structure, raw materials, excessive dependence on the oil industry, resulting in the lack of innovative impetus for economic development;At this time, if the international situation is severe and the demand for oil imports decreases, then Russia's foreign exchange income will greatly reduce and it may also be accompanied by a large amount of capital flight, the rouble will face a substantial depreciation.Due to the lack of internal impetus, Russia's economic growth is gradually slowing down, and its investment and consumption income is also declining. As a result, its economic development is in a severe situation, which will adversely affect its trade with other countries.

4.2 Suggestions

4.2.1 Fluctuations in international oil prices sometimes have a positive impact

Now Russia is in the period of oil war, it needs to gradually consolidate and expand its position in the international crude oil market and increases its competitiveness in the crude oil market, for example, it should compete for the share of exporting crude oil to powerful countries such as China so as to improve its energy status and achieve rapid economic development within a certain period of time.

4.2.2 Use a more flexible exchange rate policy

Russia started to implement the floating exchange rate system earlier. In most cases, considering the stability of the exchange rate, the Russian Central Bank would actively intervene in the foreign exchange market to minimize the exchange rate risks. It should try to use a more flexible exchange rate mechanism to achieve the market-oriented exchange rate mechanism and enhance the exchange rate flexibility, so as to prevent and defuse risks brought by financial factors to the country.

4.2.3 Use a more flexible exchange rate policy Adjust industrial structure and accelerate economic transformation

Russia should move faster to get rid of the resource-based development model, limit the export of national resources, actively adjust the industrial structure, and shift from relying on imports to substituting imports.In addition, the national income can be invested in the research and development of high-tech products and applications so as to enhance the scientific research level and manufacturing capacity of relevant enterprises and improve the way of national life and consumption.Investment in education and research should be increased, The country should also improve the competitiveness based on intelligence and knowledge and fundamentally foster its innovation capacity to increase its soft power.In addition, capital can be used in strategic investment, industrial resources can be integrated and gathered together to improve the efficiency of capital allocation.

References

1. Z Junling, W Weixian, *The Impact of International Oil Price and RMB Exchange Rate Fluctuations on China's Economy -- the basis of Global General Equilibrium Analysis*, Chinese Population Resources and Environment, **S1**,303-306(2017)
2. W Yu, W Da, *Selection and Comparative Study of Oil-producing Countries' Exchange rate Policies under the Impact of International Oil Prices*,Southern Finance, **12**,52-59(2017)
3. Z Siqing, W Bo and Q Lu, *Research on the Impact of International Oil Price Fluctuations on Russia's Real Effective Exchange Rate*, Price Theory and Practice, **01**,82-85(2017)
4. Z Xi, *Impact of International Crude Oil Price on Domestic Oil Price -- Based on the Pass-through Path of RMB Exchange Rate Fluctuation*, Journal of Hebei University, **06**,94-99(2011)
5. Z Yi, W Ke and J Lin, *Correlation Analysis between US Dollar Exchange Rate and International Oil Price Trend*, International Petroleum Economics, **05**, 55-63 (2014)

Investor Sentiment Index and Option Price Volatility Based on MIDAS Model: Evidence from China

Haiyan Xiao[1], *Yingxin* Hao[2], *and Sirong* Wu[3*]

[1]School of Mathematics and Economics, Hubei University of Education, 430205 Wuhan, China
[2]School of Statistics and Mathematics, Zhongnan University of Economics and Law, 430073 Wuhan China
[3]Shenzhen Shunheng Rongfeng Supply Chain Technology Co., Ltd, China

Abstract. The paper selects the transaction data of the option market and network data from June 1, 2015 to February 2, 2018. The principal component analysis is adopted to construct investor sentiment index. The OLS and MIDAS model are employed to study the influence of investor sentiment index on the option implied volatility in Shanghai Stock Exchange 50ETF. The empirical results show that the MIDAS model with more high-frequency information has stronger interpretation ability than the same frequency model. Investor sentiment index based on traditional indicators has a negative effect on the option implied volatility while the excessive attention of investors on the Internet would exert positive pressure on the option market. The conclusion can well explain the inherent mechanism of investor sentiment affecting option implied volatility. Therefore, it is of great practical significance to study the influence of investor sentiment on the option price volatility in China

1 Introduction

Investor sentiment is a kind of emotion formed by investors in view of their perception of various information. It contains the systematic deviation of investors' expectation of the future so that it affects the sustainability of the security market. As an active representative in China's financial market, the implied volatility of the option market is formed by the trading behavior of the market, and represents the expectation of investors for the future price volatility. It plays a very important guiding role in reflecting the pricing and optimization of options, the allocation of option assets and other regulatory policies. Under such a circumstance, it is of great importance to study the interaction between investor sentiment and implied volatility, especially to the risk prevention and sustainable development of the security market.

Bing Han [1] studied the impact of investor sentiment changes on the S&P500 option price. Zhang and Lin [2] found that there was a positive nonlinear spillover effect between investor sentiment and stock market return rate. Xiong et al. [3] found that in the market

* Corresponding author: wusirong_sharon@163.com

environment with and without trading restrictions, investor sentiment had opposite effects on the market risk management ability of CSI 300 stock index futures, but different influence depth on the price discovery ability. It is found that there are few studies on the relationship between investor sentiment and implied volatility, and in the above empirical studies, all the empirical models adopted have a common feature: the data frequency of the explanatory variable and the explained variable in the model is the same, which cannot be satisfied in practice. The traditional processing method is to convert the high-frequency data to the low-frequency data by calculating the average value of the data, or to convert the low-frequency data to the high-frequency data by interpolation method. However, it can lead to the loss of information or data distortion, which may eventually result in the reduction of research credibility.

For a more comprehensive study of the impact of investor sentiment on option market volatility, there are three contributions to the paper. Firstly, unlike prevailing studies which employed transaction data related to the security market to construct investor sentiment index, we also introduced Baidu Search Index as a novel proxy sentiment index. Secondly, as for keywords extracting, which is different from previous research [4], the keywords were only related to several specific asset names. We collected information from posts and comments from China's large investment social media Oriental Fortune by using web scraping technology, then obtained the keywords by performing text analysis. This procedure could measure the investor sentiment in a more comprehensively way. Last but not least, the mixed data model MIDAS was used to analyze the impact of investor sentiment on the implied volatility of option. The paper constructed daily, weekly and monthly investor sentiment indexes, and the effects of mixed sentiment on the implied volatility of options can be studied on a more microscopic time scale.

2 Index system

2.1 Option implied volatility

Implied volatility refers to the volatility implied by option prices observed in the market. Fleming J et al. [5] studied the volatility index of the options derived from the S&P100 index, which has been proved that can better reflect investors' expectations of the option market. China iVIX index, issued by the Shanghai Stock Exchange, is used to measure the volatility expectations of the Shanghai 50ETF over the next 30 days. It was first announced on June 1,2015 and stopped publishing on February 13, 2018. Due to the clear negative correlation between China iVIX index and the market index, it can be referred as the "panic index" of investor [6]. Therefore, the paper employ the data of China iVIX index closing price to describe the implied volatility of SSE 50ETF option market.

2.2 Investor sentiment index

2.2.1 Construct investor sentiment index based on traditional index

Traditional methods usually employ indirect indicators which include the transaction data related to the stock market to construct investor sentiment index. Since the investment decisions of these stock market transaction data are made by investors based on their psychological characteristics, there always seems to be some inherent correlation that can indirectly reflect investor sentiment. Closed-end fund can be adopted as the investor sentiment index [7,8]. Compared with a single indicator, a composite sentiment index can quantify the sentiment of option investors in a more comprehensive and multi-angle

https://doi.org/10.1051/shsconf/20219604007

perspective. Therefore, some researchers adopted the composite indicators. Baker and Wurgler [9] defined sentiment based on the closed-end fund discount, NYSE share turnover, the number and average first-day returns on IPOs, the equity share in new issues and the dividend premium. Later, Kim et al. [10] also adopted this index.

In order to construct a composite index, three representative indicators are selected as the original variables in the Chinese options market: the option contract volume, the number of open interests and the put/call trading volume ratio. The principal component analysis method used by Baker and Wurgler [9] in constructing the sentiment will be adopted. The option contract volume, the number of open interests and the put/call trading volume ratio are marked as x_1, x_2, x_3. Standardize the data before using this method. In order to eliminate the influence of variance, we extract the principal components from the correlation coefficient matrix. The composite index is constructed by the first principal component which explains 73.69% of the (standardized) sample variance. Hence, we obtain a daily sentiment index based on traditional indicators. As in Eq. 1:

$$index_{TD} = 0.6073x_1 + 0.5911x_2 + 0.5313x_3. \qquad (1)$$

We now use the Fig. 1 to illustrate the rationality of the investor sentiment composite index. The time-series diagrams of all variables during the sample period are shown in Fig. 1. We can conclude that there exists a highly correlation between the three variables and China iVIX index, which makes support to the construction of the traditional investor sentiment index. It is found that the traditional investor sentiment index is negative correlated with the option implied volatility from Fig. 1(d).

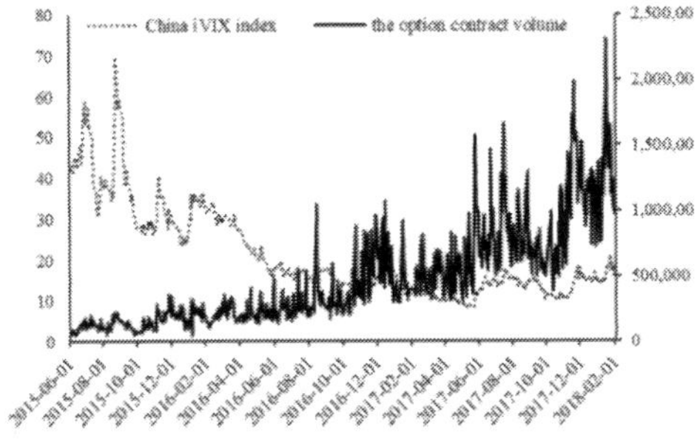

(a) China iVIX index and the option volume.

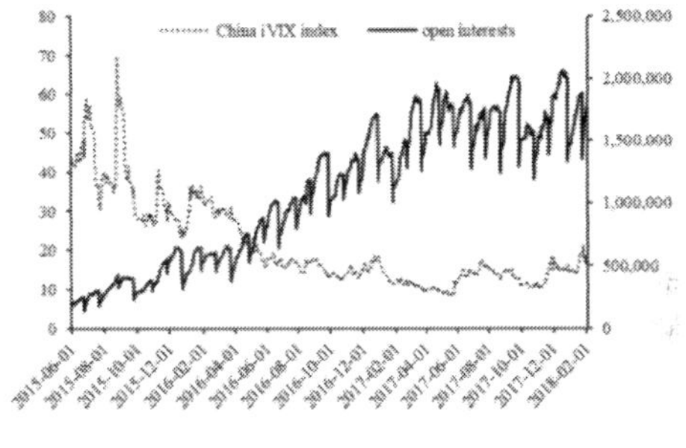

(b) China iVIX index and open interests.

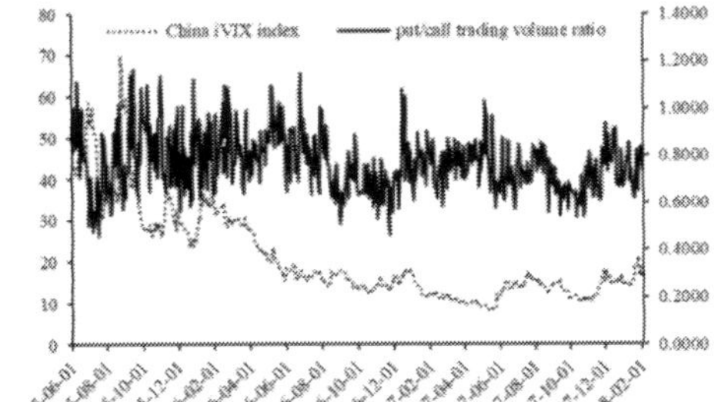

(c) China iVIX index and put/call volume.

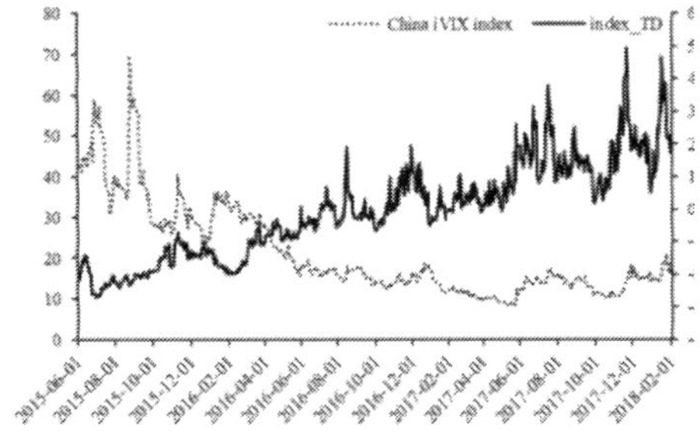

(d) China iVIX index and $index_{TD}$.

Fig. 1. Proxy variables, $index_{TD}$ and the China iVIX index.

2.2.2 Construct investor sentiment index based on Baidu index

The rapid development of Internet has provided new support for investor psychology and behavior researches. Investors are more willing to search information or express their opinions and attitudes on the online social platforms. Therefore, adopting the social media data to measure investor sentiment index can provide a better explanation for the impact of investors in the stock market.

Some researchers used microblogging forums tweets including Twitter and Weibo to show the relationship between public sentiment and the stock market. Kim et al. [11] collected text messages on the Yahoo! and found investor sentiment had predictive ability for stock returns. Wang et al. [12] was the first attempt to use Baidu Search Index for research. Therefore, using public opinion information and network information as the source of emotional information has become a common method, which has been verified that it can reflect emotions of the investors directly [4].

Scraped the text data of the postings and comments of Shanghai Stock Exchange 50ETF (short for SSE 50ETF) Option Bar from Oriental Fortune website, which include 2,326 comments from June 1, 2015 to February 2, 2018. The text information of the posts and comments crawled from the Oriental Fortune Website can extract the most frequently spoken words by shareholders, which are exactly the investors most concerned about and are most closely connected with the option market.

Keywords were selected based on the principle of frequency greater than 100 and included in Baidu Index.These 13 proxy variables including option, 50ETF, call option, market, trading, market trend, Shanghai Stock Exchange, index, buy, account opening, account, threshold and short are assigned labels as $z_1,L\ ,z_{13}$. Same as constructing traditional investor sentiment index, adopting the principal component analysis method. The first three principal components f_1, f_2, f_3 can total explain 71.8% of the (standardized) sample variance, as in Eq. 2:

$$\begin{aligned} f_1 &= 0.2913z_1 + 0.3673z_2 + L\ + 0.0008z_{13} \\ f_2 &= -0.2278z_1 - 0.2643z_2 + L\ + 0.4355z_{13} \\ f_3 &= -0.0130z_1 + 0.0157z_2 + L\ - 0.2286z_{13}. \end{aligned} \quad (2)$$

Thus, a new composite index can be constructed by the weighted sum of the first three principal components, where the weights are chosen as the variance contribution rate, as in Eq. 3:

$$index_{BD} = 44.67\% f_1 + 18.15\% f_2 + 8.98\% f_3. \quad (3)$$

Fig. 2 shows a highly similar fluctuations and trends between the $index_{BD}$ and China iVIX index. Unlike the $index_{TD}$, the correlation between the $index_{BD}$ and China iVIX index is positive. Therefore, it is rational to say that the $index_{BD}$ represents an overall search enthusiasm on SSE 50ETF options.

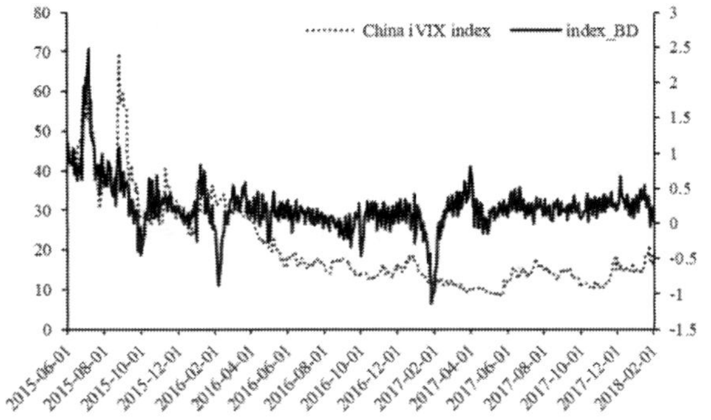

Fig. 2. China iVIX index and $index_{BD}$.

3 Methodology

Ghysels et al. [13] proposed the MIDAS model based on the distributed lag model, which can effectively study the actual reflection of high-frequency explanatory variables on low-frequency interpreted variables and avoid information loss or data distortion. The expression of the univariate MIDAS model is as Eq. 4:

$$y_t = \alpha + \beta W\left(L^{\frac{1}{m}};\theta\right)x_t^{(m)} + \varepsilon_t,\ \ t = 1,\ldots,T, \tag{4}$$

where m is the frequency multiple difference between the high-frequency data and the low-frequency data. $x_t^{(m)}$ means the m samples are taken from time $t-1$ to time t. Parameter β represents the overall influence of explanatory variables $x_t^{(m)}$ on y_t. ε_t is the error term, and T represents the entire period of y_t. $W\left(L^{\frac{1}{m}};\theta\right)$ is a polynomial weight function dependent on the parameters K and θ, the specific expression is shown in Eq. 5.

$$W\left(L^{\frac{1}{m}},\theta\right) = \sum_{k=0}^{K} B(k;\theta)L^{\frac{k}{m}}, \tag{5}$$

where $L^{\frac{1}{m}}$ is the lag operator of $x_t^{(m)}$, and $L^{\frac{k}{m}}x_t^{(m)} = x_{t-k/m}^{(m)}, k = 0,1,\ldots,K$. K is the maximum lag order of the high-frequency variable.

The MIDAS model involves the selection of polynomial weighting function and how to select parameter vector θ and lag order K in polynomial weighting function. Liu et al. [14] pointed out that Almon polynomial functions are mostly used in the prediction and analysis of Financial market volatility, while exponential Almon polynomial functions and Beta polynomial functions are mostly used in the analysis and prediction of macroeconomic. In the paper, Almon polynomial function is selected as the weight function in Eq. 6.

$$B(k;\theta)=\frac{\theta_1 k+\theta_2 k^2+\text{L}+\theta_p k^p}{\sum_{k=1}^{K}\left(\theta_1 k+\theta_2 k^2+\text{L}+\theta_p k^p\right)}. \tag{6}$$

For the choice of parameter vectors θ, we employ $p=2$ following Clements and Galvao [15]. For the selection of the lag order K, Gao and Yang [16] compared the adjust goodness of fit and significance level of parameter estimation with different lag orders. In the paper we select the optimal lag order according to the significance level of parameter estimation. Therefore, the model established in the paper is Eq. 7.

$$V_t=\alpha+\beta_1 B\left(L^{\frac{1}{m}};\theta_1\right)index_{TD_t}^{(m)}+\beta_2 B\left(L^{\frac{1}{m}};\theta_2\right)index_{BD_t}^{(m)}+\varepsilon_t, \tag{7}$$

where V_t is the closing price of China iVIX index. $index_{TD}$ and $index_{BD}$ are investor sentiment index.

4 Empirical analysis

The paper mainly studies two cases: (1) whether the investor sentiment index has a significant impact on option implied volatility and (2) whether the MIDAS model has a strong interpretation ability. To ensure that there is no spurious regression, we have to testify the stationarity of time series data before the empirical analysis. The result shows that all variables have passed the ADF test. In order to measure the pros and cons of the interpretation ability of the MIDAS model, Clements and Galvao [15] compared and analyzed the MIDAS model with the benchmark model which is OLS method under the same frequency data. Along this line, the parameter results of both models are given in Table 1 and Table 2 respectively.

Table 1. Estimation results of OLS.

Parameter	Monthly		Weekly	
	value	t-Statistic	value	t-Statistic
α	17.0064 ***	15.2034	17.4272 ***	24.6974
β_1	-5.4056 ***	-8.5374	-5.4299 ***	-13.4035
β_2	15.9687 ***	5.0480	15.3601 ***	8.7348
$Adj-R^2$	0.8133		0.7103	

Note: ***, significance at 1% level; **, significance at 5% level; *, significance at 10% level.

The estimation results of the benchmark model are given in Table1. Table 2 shows the estimation results of the MIDAS models with different frequencies. For both MIDAS model and OLS model, the parameter estimation value of $index_{TD}$ is significantly negative, while $index_{BD}$ is significantly positive. These results are consistent with the conclusions drawn from Fig. 1(d) and Fig. 2.

By comparing the $Adj-R^2$ of different models, we can judge the fit of the model. When the explained variable is the monthly implied volatility of options, the fitting effect of the MIDAS model using daily data for the explanatory variable is better than that of the OLS homo-frequency model. The same results are obtained when the weekly data are used as explanatory variables. This indicated that the mixed frequency model has stronger explanatory power than homo-frequency model. However, the fitting effect of Monthly/Weekly MIDAS model is lower than that of monthly OLS model with the homo-frequency. This shows that the interpretation ability of the MIDAS model is not uniformly better than the traditional regression model, but depends on factors such as the difference in frequency multiples used. Fig. 3 shows the fitting results obtained by using Monthly/Daily and Weekly/Daily data. It is intuitive to see that using high-frequency data can better simulate lowfrequency financial data.

Table 2. Estimation results of MIDAS.

Parameter	Monthly / Weekly		Monthly / Daily		Weekly / Daily	
	value	t-Statistic	value	t-Statistic	value	t-Statistic
α	6.1487	1.6529	14.7704***	11.3183	11.8794***	8.5274
β_1	-67.3223**	-2.5978	-1.4671**	-2.4705	-1.4110***	-2.7740
θ_{11}	72.8006**	2.3852	0.2536**	2.1594	0.2688*	1.9146
θ_{12}	-16.9833**	-2.2230	-0.0093**	-2.1072	-0.0088	-1.1418
K_1	3		25		17	
β_2	16.4099*	1.84498	1.2318*	1.6718	1.5493***	3.2233
θ_{21}	-14.0669*	-1.7044	-0.1595	-1.3916	-0.3237**	-2.3390
θ_{22}	2.9652*	1.8301	0.0054	1.5165	0.0233***	2.9410
K_2	4		34		16	
m	4		22		5	
$Adj - R^2$	0.7896		0.8216		0.8418	

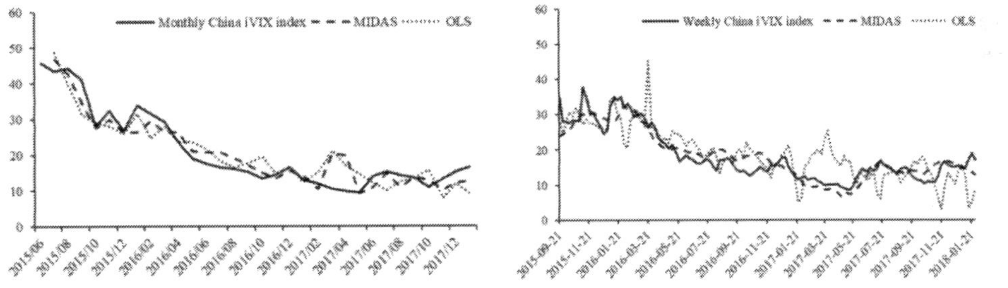

(a) fitting effect on monthly data. (b) fitting effect on weekly data.

Fig. 3. OLS and MIDAS fitting effect on China iVIX index.

It can be seen from the above empirical analysis results that Baidu Index can reflect the level of investor attention to a certain extent. If investors are interested in an option product, they will increase the search volume of relevant information, conversely, the increase in the amount of search information will also affect investors' investment expectations and behaviors, thereby affecting asset price fluctuation. Therefore, the excessive attention of investors on the Internet will exert positive pressure on the option market, which will increase the implied volatility of the option market.

In order to better verify the validity and robustness of MIDAS model, the closing price of the China iVIX index is replaced by the trading volume of SSE 50ETF. The regression results are consistent with the above results, indicating that the research results of the paper are robust.

5 Conclusions

Using the principal component analysis method, the paper constructs two investor sentiment indexes from different angles and frequencies. Two models, MIDAS model and OLS model, are established to analyze the influence of investor sentiment on the implied volatility of options. The empirical results show that the investment behavior of investors has a significant negative impact on the implied volatility of options, and investors' excessive attention to the Internet can increase the implied volatility of the options market. In the process of research, we found that the mixed frequency model has better explanatory power than the homo-frequency model.

The findings of the paper can show the internal mechanism of the influence of investor sentiment on the option implied volatility, and provide references for other relevant researchers. Meantime, as for relevant government departments, in order to ensure a stable, healthy and sustainable development of the security market, it is recommended that the government improve the information disclosure system; monitor investor sentiment changes on official platforms; reasonably guide investors to invest rationally; prevent malicious transmission of false information on social media, and prevent disruption of order in the security market.

References

1. B. Han, Rev. Financ. Stud. **21,** 387 (2008)
2. G. Zhang, Y. Lin, J. Appl. Stat. Manag. **Online,** 1 (2020)
3. X. Xiong, K. Xu, D. Shen, Syst. Eng. - Theory Pract. **40**, 2252 (2020)
4. X. Meng, X. Meng, Y. Hu, Macroecon. **1**, 144 (2016)
5. J. Fleming, B. Ostdiek, R. E. Whaley, J Futures Markets **16**, 353 (1996)
6. R. E. Whaley, J. Portfolio Manage. **26**, 12 (2000)
7. C. M. C. Lee, A. Shleifer, R. H. Thaler, J Finance **46**, 75 (1991)
8. G. W. Brown, Financ. Anal. J. **55**, 82 (1999)
9. M. Baker, J. Wurgler, J Finance **61**, 1645 (2006)
10. J. S. Kim, D. Ryu, S. W. Seo, J. Bank Financ. **42**, 166 (2014)
11. S. H. Kim, D. Kim, J. Econ. Behav. Organ. **107,** 708 (2014)
12. P. Wang, W. Zhang, X. Li, D. Shen, J. Econ. Interact. Coord. **14,** 377 (2019)
13. E. Ghysels, P. Santa-Clara, R. Valkanov, Chapel Hill: University of North Carolina, 2004
14. J. Liu, H. Liu, Z. Yin, Econ. Sci. **5**, 23 (2010)
15. M. P. Clements, B. Galvao, J. Appl. Econ. (Chichester Engl) **24**, 1187 (2009)
16. B. Gao, C. Yang, Int. Rev. Econ. Finance **49**, 69 (2017)

The Impact of Blockchain Technology on Stock Price: An Emprical Study

Minhe Xu*

Economics and Management Department, Beijing Jiaotong University, Beijing, China

Abstract. With the rapid development of blockchain technology, more and more attention has shifted from the theoretical research of blockchain technology to the application of specific business of the company. However, there are few literatures on the quantitative research on the impact of blockchain technology on the company's market value after it is applied to the company's actual business. In this paper, 73 listed companies published blockchain application announcements from 2016 to 2019 are selected as the research objects, and the short-term event analysis method is used to quantitatively analyze the impact of the application about blockchain technology on the market performance of listed companies. The results show that: after the announcement of blockchain application in their own business, listed companies have a more significant abnormal income, which shows that the capital market has a significant positive response to the application of blockchain. Finally, this paper puts forward the theoretical and practical significance, limitations and future research.

1 Introduction

At present,The rise of blockchain technology has brought new development opportunities to many fields, which is both an opportunity and a challenge for China. In recent years, the academic research on the theory and application of blockchain has increased rapidly, especially in China. Various relevant departments of the country have gradually introduced various policies to encourage the development of blockchain. Enterprises from all walks of life have begun to combine blockchain technology with their own business through their own research or cooperation with technology companies, so as to give full play to the advantages of blockchain. However, there is no quantitative analysis on the market reaction of Chinese companies after applying this technology in their own business. This paper is based on this background. In this paper, 73 listed companies published blockchain application announcements from 2016 to 2019 are selected as the research objects, and the short-term event analysis method is used to quantitatively analyze the impact of the application of blockchain technology on the market performance of listed companies.

* Corresponding author: 19125563@bjtu.edu.cn

2 Literature review

Since the advent of bitcoin in 2008, blockchain technology, as the underlying core foundation, has been highly concerned by political, industrial and academic circles. Many scholars study blockchain technology from different aspects of theory and practice, and are committed to applying blockchain technology to all walks of life, especially in China, from the financial field to education, supply chain, logistics and other fields [1, 2]. Then, for the impact of the application of emerging technologies on the value of the company's stock has been studied by scholars, using the methods of event analysis, regression analysis and so on. The application of event analysis quantifies the impact of the emergence or application of new technologies on the company's stock price. For example, some scholars empirically test the impact of the implementation of 3D printing (3DP), cloud computing and intelligent logistics policies on the stock return rate; there are also studies on the price response of listed companies to blockchain related announcements[3];There are also scholars who study the impact of "blockchain" or "bitcoin" added to the name list of companies on abnormal returns, so as to judge the stock market reaction[4, 5].Therefore, this paper draws on the experience of the above literature to study the response of the capital market to the specific application of blockchain under the background of China. We adopt the short-term event analysis method, taking the Chinese listed companies issuing blockchain application announcements as the event samples, and judge the market reaction by studying abnormal returns.

3 Hypotheses development

H1: Stock markets react positively to the announcement of blockchain application in Chinese Enterprises.

Whether it is the research of technology theory or the application of specific business, blockchain technology has a huge development prospect. At present, the development of blockchain technology is only the tip of the iceberg. First of all, from the advantages of blockchain technology itself, due to its traceability, symmetry and other characteristics, it will bring different degrees of benefits to the company after it is applied to the business of many industries [6]. Secondly, due to the guiding role of the policy, the government promulgates policies, which indicate the development potential of the industry, along with some supportive actions of funds and talents, which are conducive to the future development of enterprises. In recent years, national departments have successively issued relevant policies to encourage the development of blockchain technology, which also has a certain impact on the stock value of companies applying blockchain technology. Finally, when the announcement of the practical application of this technology is announced, it not only indicates that the company is in good condition, but also has a clear plan for its future development, which will strengthen investors' confidence in development. Therefore, we expect that the application of blockchain technology to the company's specific business will produce a positive market reaction.

4 Research methodology

4.1 Data collection

We collected the variables to be studied from various sources. First of all, we obtain the news and announcement of blockchain application of Chinese enterprises from the official website, Sohu, Tencent and other news aggregation channels [7]. The application of blockchain technology in China has been booming in recent years. Therefore, we collected the announcement of the company's application of blockchain in 2016-2019, which also conforms to the basic hypothesis of short-term event research. Then, by carefully reading each published news text, we only retain those announcements annouexcluding those related to research and development of blockchain technology. In additioncing that the company has actually applied blockchain technology in specific business, n, in order to make the research more targeted, we focused on the companies listed on the Chinese stock exchanges, excluding those listed on the New York Stock Exchange and NASDAQ. Finally, we cross check the same blockchain application announcement reported on different channels, and use the earliest publication date as the event date. Therefore, through a series of screening, we obtained 101 announcements.

We obtain the daily stock return and composite index of all target companies from China Securities Market and accounting research (CSMAR) database as the daily market return, so as to analyze the impact of blockchain technology announcement on stock returns by applying short-term event analysis method. CSMAR database can query the stock return data of all listed companies in China in the past 30 years.

4.2 Event study methodology

We use the short-term event research method to quantify the impact of the event of blockchain application on the abnormal return of stocks [8]. Specifically, abnormal return refers to the difference between the actual return rate of a company after it publishes the announcement on the application of blockchain technology in its business and the expected stock return assuming that no blockchain application announcement is published. According to the basic method of short-term event research, we use the market model to estimate the expected stock return when the company does not publish the blockchain application announcement.

$$R_{it} = \alpha_i + \beta_i R_{mt} + \varepsilon_{it} \tag{1}$$

$$E(R_{it}) = \hat{\alpha} i + \hat{\beta}_i \, R_{mt} \tag{2}$$

$$AR_{it} = R_{it} - E(R_{it}) \tag{3}$$

$$CAR_i[-t_1, t_2] = \sum_{t=-t_1}^{t_2} AR_{it} \tag{4}$$

The specific estimation process is as follows. First, we take 11 days to 100 days before the announcement date of the company's release of blockchain applications as the estimation period, so as to avoid affecting the estimation of the expected stock return rate of the announcement. Then, according to the formula (1), the daily stock return (R_{it}; the return rate of the company R_i on t day) and the market return (R_{mt}; corresponding to the return rate of the stock exchange's composite index on t day) during the estimation period are regressed.The estimated values of α_i and β_i are obtained by regression calculation, and then according to the formula(2), we calculate the expected stock return[$E(R_{it})$]during the event period. Next, calculate the abnormal return (AR_{it}) caused by the announcement of the blockchain application, that is, according to the formula (3), the difference between the actual stock return (R_{it}) and the expected stock return [$E(R_{it})$]. In addition, the sample number of event study is reduced from 101 to 73, because some sample companies are excluded because of the lack of all stock return data in the estimation period or the event period, or the regression effect of the company return and market return is not significant in the estimation period.

According to the previous research on short-term events, we selected three trading days before and after the event (i.e. day - 1 to day 1) as the event window to explain the possible information leakage before the event and the impact after the announcement was issued. In order to observe the total effect of the event, we also calculate the cumulative abnormal returns (CARs) in the event window, according to the formula (4). Consistent with previous event studies, we used parametric test (t-test) to analyze the statistical significance of ARS and CARs. Table 1 shows the average value, median value and parameter test (i.e. t test) of daily excess return and cumulative abnormal return in event window by event window and model, and tests whether the average and median of daily abnormal return and cumulative abnormal return are significantly different from zero.

5 Data analysis and results

The abnormal return test results of short-term event analysis are shown in Table 1. Panel A in Table 1 shows the abnormal returns on the release date of the blockchain application announcement and the days before and after the announcement (i.e., day - 1, day 0 and day 1). We found that the abnormal returns on day -1 were not statistically significant ($P > 0.1$ in t-test) on day - 1, but they became significant on day 0 and day 1 ($P < 0.1$ in t-test). This shows that there is no sign of information leakage before the listed companies issue the blockchain application announcement, excluding the possibility of issuing the announcement after the stock market is closed. In addition, the average rate of return and median of Chinese enterprises to the announcement are positive, indicating that Chinese enterprises have a positive reaction to the practical application of blockchain.

Table 1. The test results of event study.

Panel A: Abnormal Returns (ARs)				
Day	**N**	**Mean ARs**	**Median ARs**	**t-Test**
-1	73	0.00038821	-0.00009219	0.208
0	73	0.008554928	0.001561338	2.632***
1	73	0.007037378	0.00246896	2.356**
Panel B: Cumulative Abnormal Returns (CARs)				
Days	**N**	**Mean CARs**	**Median CARs**	**t-Test**
(-1,0)	73	0.008943137	0.00070318	2.228**
(0,+1)	73	0.015592306	0.002802485	2.89***
(-1,+1)	73	0.015980515	0.000938536	2.616**

* p<0.10 (two-tailed tests). ** p<0.05 (two-tailed tests). *** p<0.01 (two-tailed tests).

The test results of cumulative abnormal returns (CARs) shown in panel B further prove that the average and median values of cumulative abnormal returns (i.e. - 1 to 0, 0 to 1, and - 1 to 1) in the event window are positive, while the results of t test ($P < 0.05$) are significant. Therefore, H1 is supported.

6 Discussion and conclusion

From the analysis results, the announcement of the application of blockchain in China's listed companies' own business will have an immediate significant positive impact on the stock market of enterprises. From the theoretical aspect, this study deepens the understanding of the enterprise value when the blockchain technology is applied to the company's specific business. From a practical point of view, we will suggest that enterprises can invest and use blockchain technology according to their own business conditions. Finally, although our whole research process is as objective and comprehensive as possible, there are still some limitations and areas for further study. The results of this paper mainly analyze the real-time market reaction based on the short-term, but not the medium and long-term impact. Therefore, we hope that future researchers can further expand the sample size and improve the representativeness; consider the medium and long-term impact of blockchain application on the company value; and study the impact of blockchain technology on different industries according to the industry scope.

References

1. Q. Wang, M. Su, R. Li, J. Clean. Prod. **264**, 121742 (2020)
2. L. Hughes, Y. K. Dwivedi, S. K. Misra, N. P. Rana, V. Raghavan, V. Akella, Int. J. Inform. Manage. **49**, 114-129 (2019)
3. C. Daniel, D. G. Baur, L. Zhangxin, Y. Joey, SSRN Electronic Journal. (2018)
4. P. Sharma, S. Paul, S. Sharma, Econ. Lett. **186**(Jan.), 108818.1-108818.3 (2020)
5. A. J. A, C. J. B, Econ. Lett. **181**, 178-181 (2019)
6. D. D. F. Maesa, P. Mori,J. Parallel. Distr. Com. **138**, 99-114 (2020)
7. H. K. S. Lam, Y. Zhan,M. Zhang, Y. Wang, A. Lyons, Int. J. Prod. Econ. (2019)
8. L. Ding, H. K. S. Lam, T. C. E. Cheng, H. Zhou, Int. J. Prod. Econ. **200**(JUN.), 329-342 (2018)

Research on the Influence of ODM Mode on Consumer Decision

Ming Tian*,

School of Economics and Management, Beijing Jiaotong University, Beijing, China

Abstract. As an emerging new consumption model, the ODM e-commerce model accurately locates consumers and uses consumers' shopping needs as the marketing strategy of the e-commerce platform. This is a huge impact on traditional physical sales and existing e-commerce models. In order to explore the relationship between ODM model products and consumer decision-making, this article takes the NetEase selected represented by ODM e-commerce model as the research object. The results found that there is a significant positive correlation between the types of commodities and ODM model and consumer decision-making. Consumer participation has a positive regulatory effect on the relationship between ODM model and consumer decision-making. The research conclusions provide theoretical support for the ODM e-commerce model to increase user stickiness, increase consumer decision-making weight, and expand marketing models.

1 Introduction

In recent years, our country's e-commerce industry has developed vigorously, and various types of e-commerce websites have sprung up, greatly enriching consumers' shopping methods and satisfying the shopping needs of different consumers. With the continuous development and growth of the e-commerce industry, a series of new e-commerce sales models have been born. Consumers have also experienced various vertical e-commerce model changes from C2C to B2C to ODM new consumption. Online shopping gradually become the new normal in the retail industry, especially the ODM e-commerce model represented by NetEase selection. Once it emerged, it quickly became a new consumption model leading other types of e-commerce platforms and grew into the most influential vertical e-commerce brand at this stage.

2 The background and significance of the ODM model

With the continuous development of social economy, people's living standards are rapidly improving, and more and more people are entering the new middle class. Consumers have a higher pursuit of product style and quality, our country is in a new consumption era of consumption upgrades. NetEase CEO Ding Lei mentioned: our country's consumer groups are undergoing an evolution. In the past, everyone was pursuing how to buy, but now

* Corresponding author: 19125548@bjtu.edu.cn

everyone is pursuing how to buy better and more tasteful. This phenomenon is called new consumption, and it focuses on the changes in consumers' consumption view and consumption behavior.

Simultaneously with the emergence of new consumer concepts, the e-commerce traffic dividend period has faded, and it has begun to enter a bottleneck period and a transition period when the growth rate is gradually slowing down, the structure is relatively solidified, and the competition is significantly intensified. For early consumers, online shopping is a new thing. Compared with offline shopping, the emergence of e-commerce enables consumers to choose a variety of products online without leaving home. The C2C e-commerce model by TaoBao has come into being, covering all aspects of life, and satisfying consumers' willingness to shop conveniently and quickly. When consumers are used to the e-commerce shopping model, businesses will not hesitate to sell fakes in order to get higher profits. Ensuring that the purchase of genuine products on the e-commerce platform has become a topic that consumers really care about. In this context, the B2C e-commerce model represented by JD.com is brand-oriented and caters to consumers' demand for authentic products.

The global market research company Euromonitor released a report on the top ten global consumer trends in 2020, pointing out that with the increase in purchasing power, consumers are pushing brands to accept their values, that is, consumers refuse mass production and ordinary products, and tend to simple and return to basic needs[1]. Consumers expect truly differentiated products and experiences[2]. At the same time, under the guidance of the supply-side reform policy, our country's manufacturing industry is also accelerating its transformation and upgrading[3]. Many domestic manufacturers have already had the production of higher-quality products and certain design capabilities. In this economic context, the ODM e-commerce model represented by NetEase selected is more in line with the psychological needs of modern consumers. This trend reflects the transformation of the power relationship between sellers and consumers. In the past, consumers had to rely on specific brands or information sources to buy the goods they needed. Now, companies must continue to innovate, lower prices, improve and beautify products can attract consumers.

3 Brief description of the ODM model

With the development of China's manufacturing industry, it has been possible to produce high-quality products with first-class equipment, production technology, management capabilities and technical personnel, and the foundry has become a skilled factory in China. The foundry is a manufacturer in the OEM model. Its full name is Original Equipment Manufacturer. However, with the rapid development of OEMs in India, Vietnam and other countries, low-cost labor changes in the international situation, China's manufacturing industry is facing the problem of reduced foreign trade orders, overcapacity, and the continuous development of the domestic economy[4]. People are no longer just the pursuit of low-cost products, but more of the high-quality products, so the foundry is facing an extremely urgent need for transformation[5].

At present, there are two main channels for factory transformation. One is to create its own brand to realize the OBM production mode. Its full name is Original Brand Manufacturer, which means that the factory operates its own brand, or the manufacturer creates its own products, produces and sells products with independent brands. Since OEMs need to have a complete marketing network to support OBM, the effort and cost required are much higher than OEMs, and their own brands will conflict with their own OEM customers, few OEMs can transform to create its own brand.

The other is to realize the ODM production mode. The full name is Original Design Manufacturer, which means that after a manufacturer designs a certain product, it may be favored by other companies, require the latter's brand to be used for production, or slightly modify the design to produce. Among them, the manufacturer that undertakes the design and manufacturing business is called the ODM manufacturer, and the product it produces is the ODM product[6]. Different from traditional e-commerce, the ODM model goes beyond intermediate channel providers and directly connects to high-quality upstream manufacturers. It goes deep into the entire supply chain of product selection, production, sales, logistics, and after-sales. It follows the pricing rule of "cost price + value-added tax + postage", eliminating high brand premiums, advertising, and intermediary costs, and achieving the "de-branding" of high-quality products[7]. Coupled with the impact of the COVID-19 this year, many sales have moved from offline to online. It is also difficult for e-commerce companies to control the construction period and quality of foundry products. Therefore, under the special economic background, with the high-quality and low-price service concept, it can attract more consumers, have a large number of audiences, and better reflect its characteristics.

Once the ODM e-commerce model represented by NetEase selected has emerged, it has quickly become a new consumption model that leads other e-commerce platforms, such as Taobao Xinxuan and JD.made, which have had a huge impact on the traditional e-commerce operation model. The mass foundry mode of production is gradually replaced by small batches of customized production. Small and refined products are more in line with the consumption concept of the modern new middle class, and there is a larger consumer market.

The ODM e-commerce model focuses on differentiated marketing strategies. Under the background of new consumption upgrades, our country's manufacturing industry accelerates the pace of supply-side reform, the expansion of enterprises stimulates domestic demand, and reduces consumers' growing needs for a better life, as well as unbalanced and insufficient development. The contradiction between the two has major strategic influence and significance. Existing research on ODM models mostly studies the operation management and development strategies of ODM models from the upstream perspective, that is, the supply chain perspective, and rarely studies the ODM e-commerce models from the downstream perspective, that is, the consumer perspective. In order to have a deeper understanding of the influence mechanism of ODM e-commerce model on consumer decision-making, this article uses NetEase selected of e-commerce platform as the research object to explore the influence mechanism of ODM e-commerce model on consumer decision-making under the influence of new consumption concepts. Provide a modest force for the optimization and development of the ODM e-commerce model.

4 NetEase selected overview

As NetEase's original life-quality e-commerce platform, NetEase selected is the first e-commerce company to adopt the ODM model, focusing on home life and other products. Advocate the business philosophy of "good life, not so expensive", connect directly with famous manufacturers, eliminate brand premiums and intermediate links, and select high-quality, high-quality products for consumers. NetEase selected positions its audience as the new middle class. Most consumers are between 20 and 35 years old. This part of consumers is characterized by limited income but pursuing high-quality life. The first choice for consumption is rational consumption that focus on product content and tend to products with high quality and low price that improve the quality of life[8]. In order to meet the needs of consumers, NetEase selected has always adhered to the standard of excellence in product selection, saving consumers time and cost of purchasing products.

The entire process from production to sales of the products on the NetEase selected platform is a standard ODM operation process. Therefore, this article uses the NetEase selected as the research object. In this platform, each product can be collected from the shelves, views, and collections. Inventory, sales volume, repurchase rate and other information, through the development of data sets, data analysis can be used to analyze the impact of ODM e-commerce model products on consumer shopping decisions.

5 The influence of ODM mode on consumer decision-making

5.1 Data Collection

Use a Java-based web search engine to collect data from the official website of NetEase selected, and use commodities as the unit of data collection. From the official website, you can see that NetEase selected includes eight categories, such as home life, clothing, shoes and bags, food and beverages, personal care, cleaning, maternal and child parenting, sports travel, digital home appliances, and strict selection of global. Different categories of goods have different effects on consumers' shopping decisions. So this article uses the commodity categories of NetEase selected as control variables to analyze. As of December 23, 2019, the NetEase selected homepage has 4,293 products. In order to improve the accuracy of data analysis, different attributes of products will be captured, such as product sales, product evaluation and other data. According to whether the product has the "NetEase self-operated brand" logo, the ordinary products and ODM model products of the NetEase selected platform are distinguished. Since the platform cannot capture the sales volume, according to the characteristics that the evaluation volume is basically consistent with the sales volume[9]. The evaluation volume is used as the data of the sales volume. The number of additional reviews based on the product is used as the analysis data of consumer participation. In addition, it also includes the category of the product. After collecting the data, the data is normalized to lay the foundation for subsequent empirical analysis. In order to avoid deviations from the analysis of the model due to the type of commodities, the commodity category is analyzed as a control variable.

5.2 Define research variables

In order to analyze the impact of the ODM model on consumer decision-making, this paper analyzes the collected data based on the SPSS software platform. The product category is used as the control variable, the ODM model product is used as the independent variable, and the product sales volume is the dependent variable. Consumer participation is used as a moderating variable. The specific measurement methods of the variables are shown in Table 1.

Table 1. Variable measurement.

Variable		Description	Measurement
Independent Variable	ODM	Whether the product is self owned brand of NetEase.	0: Non self owned brand of NetEase 1: Self owned brand of NetEase
Dependent Variable	Sales volume of goods	The quantity of productin sold.	evaluation quantity
Moderator Variable	Consumer continued participation	Whether the consumer has any additional comments after purchasing the product.	0: No additional comments 1: Having additional comments
Control variable	FMCG (Fast Moving Consumer Goods)	Whether the product is FMCG type	0: Not FMCG 1: FMCG

ODM: This variable is Whether the product is self-owned brand of NetEase. Use "NetEase self-operated brand" as the distinguishing mark of this variable, and code the logo with "NetEase self-operated brand" in the crawled product data as "1", and code "0" without this logo.

Sales volume of goods: This variable is the quantitative data representation of the model. There is no merchandise sales volume displayed on this platform. According to the default evaluation and content evaluation in the evaluation volume, it can be basically assumed that the evaluation volume is consistent with the merchandise sales volume data.

Consumer continued participation: This variable is the expression of consumer participation in the product after receiving the product. Use the additional comment of the product as the variable. If there is an additional comment, the product is coded as "1", and if there is no additional comment, the code is "0".

Control variable: The product category is used as control variables. Due to the relatively low threshold of ODM for apparel, footwear and bags, many businesses will form an ODM model to increase profits. In order to reduce the impact, it will be used as the control variables.

5.3 Descriptive statistical analysis

From the perspective of the overall commodity attributes of NetEase selected, ODM model products account for the vast majority (78.9%), and there are still some non-ODM model products (21.1%). Among them, home life, food and beverages, clothing, shoes and hats, and selected Global's products on the NetEase selected platform accounted for 16.1%, 13.4%, 19.1% and 26.1% of the total number of products respectively. Since NetEase selected adjusts its operation and sales plans according to consumers' needs and feedback, the quantity ratio of product categories and ODM products in each product category also confirm the positioning of NetEase selected for consumers. New middle-class consumers who have a certain spending power and tend to have a higher quality of life. They are generally between 25 and 35 years old, and they pay more attention to quality life and healthy life.

In the survey of consumer participation, it was found that among the products on the NetEase selected, products with review accounted for 74.9% of the total number of products, of which ODM products accounted for 82.7% of the total number of products with review. Commodities accounted for 25.1% of the total number of commodities. On the whole, the high level of consumer participation on the NetEase selected indicates that users are more sticky and the shopping platform is more active. From the product categories, except for the strict selection of the global product category, the number of products reviewed for the other seven categories accounted for more than 80% of the total number of products in each category. ODM products also accounted for the vast majority, indicating that NetEase consumers on the selected platform are willing to share their willingness to use with other consumers, which also helps the platform increase the weight of consumers' decision-making when choosing products, which has a certain impact on consumers' shopping choices.

5.4 Correlation analysis

Use SPSS software to analyze the correlation between ODM model products and consumer decision-making, use product categories as control variables, ODM products as independent variables, product sales as dependent variables, and consumer participation as a moderating variable to construct a model. The method of stepwise hierarchical regression is adopted. Model 1 is the correlation analysis between control variables and consumer decision-making; Model 2 adds independent variables to analyze the correlation between ODM model goods and consumer decision-making; Model 3 adds moderating variables to analyze consumer participation in ODM model The impact of the relationship between goods and consumer decisions. The results of stepwise hierarchical regression analysis are shown in Table 2.

Table 2. Hierarchical regression results.

Variable	Model 1	Model 2	Model 3
Dependent variable: Sales volume of goods			
Control variable			
FMCG	0.250***	0.253***	0.229***
Independent variable			
ODM		0.159***	0.145***
Moderator effect			
Consumer continued participation * ODM products			0.181***
Constant	1.406E-15	1.298E-15	-0.032
R-square	0.075	0.100	0.130
Observations	3474		

Note: *p < .10; **p < .05; ***p < .01. All tests are tailed.

The correlation analysis results show that in Model 1, the product category and consumer decision-making show a positive correlation (β=0.250***); in Model 2, the independent variables are added to the model, and the correlation analysis results show that ODM There is a significant positive correlation between model products and consumer decision-making (β=0.159***). As the control variable of product category, its correlation with consumer decision-making has no significant change; in Model 3, the consumer participation As a moderating variable, the analysis results show that consumer participation has a positive regulatory effect on the relationship between ODM model products and consumer decision-making (β=0.181***). The closer the relationship between ODM model products and consumer decision-making.

6 Conclusion

This paper takes the NetEase selected platform represented by the ODM e-commerce model as the research object, and explores the relationship between ODM model products and consumer decision-making. The research found that there is a significant positive correlation between ODM model products and consumer decision-making. There is also a significant positive correlation between consumer decisions, and consumer participation has a positive regulatory effect on the relationship between ODM model products and consumer decisions. The research conclusions provide theoretical support for the ODM e-commerce model to increase user stickiness, increase consumer decision-making weight, and expand marketing models.

References

1. Ten global consumption trends in 2020, [online] Available: http://xue.ecduo.cn/article-78899.html
2. Q. Wang, The Effect and Mechanism of "Internet+" Promoting my country's Consumption Upgrade, Series of Finance and Economics: Journal of Zhejiang University of Finance and Economics, **12**, (2016).
3. M. Le, Wiley, Creating Strategic Leverage: Matching Company Strengths with Market Opportunities, Wiley, **9**, 4, (1992).
4. Q. Luo, The Dilemma and Countermeasures of China's Manufacturing Industry Development, 2nd International Conference on Contemporary Education and Humanities and Social Sciences, (2017).
5. Davies, Antony , and T. W. Cline, A consumer behavior approach to modeling monopolistic competition, Innovar Revista De Ciencias Administrativas Y Sociales **26**(6), 797-826, (2006).
6. R. Chen, Analysis of NetEase's strict selection model based on the new consumption era, Market Modernization, **15**, (2018).
7. T. Q. Hao, Z. Z. Ma, Z. H. Yang, et. al ODM model quality e-commerce under consumption upgrades-taking NetEase's strict selection as an example, Economic and Trade Practice, **230**, 12, (2018).
8. Y. Ling, Computer Science in Industrial Application, (2015).
9. N Hu, L Liu, J. J Zhang, Do online reviews affect product sales? The role of reviewer characteristics and temporal effects, Information Technology & Management, **9**, 2008.

Study on Post-editing for Machine Translation of Railway Engineering Texts

Yuting Li and *Xiuying* Lu*

Research Center for Applied Translation of Transportation and Engineering, School of Foreign Languages, East China Jiaotong University, Nanchang 330013, P.R.China

Abstract. With rapid development of China's railways, there are more overseas construction projects and technical exchanges in the field of railway engineering, which have generated widespread demands for translation. To meet the increasingly growing demands for translation of railway engineering texts, the mode of machine translation plus post editing (MTPE) has been frequently applied besides traditional human translation (HT) for the combination of translation quality and efficiency. Through the case study of post editing in the machine translation of China's High-Speed Railway by Google Translate, this paper discusses the common error types of machine translation in railway engineering translation, and puts forward corresponding post editing strategies, so as to provide references for MTPE of railway engineering translation in the future. It is hoped that research on post editing in the mode of machine translation for railway engineering texts may improve the translation quality and efficiency, thus helping speed up the process of China's railway going global.
Keywords: Railway engineering texts, Machine translation, Post editing, strategies.

1 Introduction

Since the first railway was built and put into use in Britain in 1825, railway has gradually become an important mode of transportation in human society and has made great contributions to the development of human society and economy. According to the *Initiative of Vision and Action on Jointly Building Silk Road Economic Belt and 21st-Century Maritime Silk Road* issued in March 2015, the connectivity of infrastructure has been listed as the priority under the Belt and Road Initiative. As the significant part of the Belt and Road Initiative, China's railway, especially high-speed railway has ushered in a golden opportunity. Railway construction is one of the infrastructure industries that has developed rapidly in China in recent years, and many projects have been contracted abroad. The overseas exchange of railway technologies has been increasingly frequent. Demands for related translation are on the rise, including the translation of code for design and construction of railway, project tenders, training of professionals, related academic works etc.

The process of "China's railway going global" requires active participation of translation, during which, the mode of machine translation plus post editing (MTPE) has been

* Corresponding author: luxy0203@163.com

increasingly involved in. FENG Quangong and ZHANG Huiyu [1] pointed out that "with the development of machine translation technology and the increase of translation demand, the role of machine translation plus post editing in the language service industry has been widely recognized, especially by some large corporations". LI Mei and ZHU Ximing [2] stressed that "due to the unreliability of machine translation, post editing has become an indispensable link in machine translation to improve translation quality, which directly determines the quality, speed and cost of translation". Therefore, it can be found that post editing, as an important part of machine translation, plays a key part in improving translation quality and efficiency. To some extent, the mode of MTPE provides an efficient way for railway engineering translation.

This paper tries to focus on the post editing through the case study of machine translation for China's High-Speed Railway by Google Translate, probing into common error types of machine translation in the field of railway engineering translation, and proposing corresponding post editing strategies, so as to provide references for MTPE of railway engineering translation in terms of translation quality and efficiency.

2 Post editing in the mode of machine translation plus post editing

2.1 The mode of machine translation plus post editing

The concept of machine translation was first put forward by an American scientist Warren Weaver. Machine translation (MT) is the automatic translation of the text from one language to another or multiple natural languages by the computers [3]. It can be roughly divided into three types according to the technology used: rule-based translation, corpus-based translation and neural machine translation, among which, the most popular machine translation system is neural machine translation. It is believed that machine translation has the advantages of high speed, low cost and consistency of terminologies [4]. Although great achievements have been made in MT technology, on the whole, the quality of output of MT is still not comparable to that of HT [5]. At the 9th MT Summit IX in United States, John Hutchins [6], a famous British linguist and information scientist, pointed out that "there is no substantial improvement has been made in the quality of machine translation, and the problems unsolved 50 years ago are still exist". Hence, the mode of MTPE has been frequently applied by language service providers (LSP) for translation practice. The combination of machine translation and post editing develops rapidly, which gives full play to both the efficiency of machine translation and the quality of human translation. This mode not only satisfies the demands of the rapid developed translation market, but also promotes the improvement of translation technology, as well as the exchange and cooperation between the academic and industry. In addition, LIU Yanli [7] pointed out that the model of MTPE is very suitable for the translation of technical and practical texts, which can effectively improve the translation efficiency. Plitt and Masselot [8] found in their experiment that the mode of MTPE greatly improved the translators' productivity by 74% and the time consuming reduced by 43% on average. In addition, many research on post editing have proved that the mode of MTPE is more efficient than human translation [8-10].

2.2 Error types and procedures for post editing

The development of post editing has always been closely associated with machine translation, which is the basis of post editing. As defined by Allen [11], post editing is a process to edit, modify and correct the texts that have been translated from one language to another language by an MT system. According to CUI Qiliang [5], post editing in a broad sense is "a process in which professional editors manually review or partially revise the MT output according to the specific quality standard in order to achieve high quality and translation efficiency in an integrated translation environment". FENG Quangong and LI Jiawei [12] defined post editing as "the process of processing and modifying the original MT output according to specific purposes, including correcting translation errors and improving the accuracy and readability of MT output." This concept is more specific, and points out the specific aspects of post translation processing and revision, which has a strong guidance for the future research on the concept of post editing.

2.2.1 Error types of machine translation

CUI Qiliang and LI Wen [13] concluded 11 error types of machine translation, namely, under translation, over translation, terminology mistranslation, form errors, format errors, addition or omission of meaning, redundancy, errors in parts of speech, clause mistranslation, inappropriate word order and constraint of sentence structure. YAN Qingjia and YAN Wenpei [14] suggested that the errors produced by machine translation can be classified into several common types, which mainly come from "the errors caused by redundant words or disjunctive words, mistaken words selection, syntactic structure conversion, misuse of specific words in a certain language, and wrong forms, etc." FENG Quangong and LI Jiawei [12] concluded that the scope of the post editing mainly includes: checking whether the names of people and place and terminologies are used correctly and consistently; whether the word order needs to be adjusted; whether the word translation is accurate and unambiguous; whether it is in line with the target language syntax and expression practice; whether there are additional or missing translations; and whether there are cultural or ideological conflicts, etc.

MT can only play the role of "primary translation". To ensure the quality of translation, it's necessary to combine with the key part of post editing. Post editing of machine translation is the operation of editing and modifying the MT output, which can improve the quality and efficiency of translation to a certain extent. In today's era, with the rapid growth of information, the total amount of texts and data to be translated is beyond the competence of human translation. Therefore, as a post editor, the error types in machine translation including mistranslation, phrase collocation, conjunction, lack of subject, predicate or other components of the sentence, repetition, disagreement of subject and verb, errors in part of speech and format must be noted, for the correct and quick recognition of the errors in MT output can lead to a sound modification.

2.2.2 Competence and strategies for post editing

2.3 Error types of machine translation

The competence of post editing requires the comprehensive mastery of editing and translation, involving the strong competence of both source language and target language, recognizing of subject knowledge, application of software and cross-cultural communication. According to Kruger [15], editing ability includes the ability of reading and writing the target language, focusing on the text level and details, being highly sensitive to the text, the author, the context and readers, and the understanding and mastery of editing process and result. Therefore, strategies taken are closely related to the various capabilities for post editing, which are conducted at lexical, syntactic and discourse levels by linguistic analysis, semantic understanding, terminology management, error recognition and checking, stylistic refinement, etc.

Railway engineering translation is a kind of translation demand arising with the acceleration of "China's railway going global". The application of MTPE in the practice of railway engineering translation will greatly improve the translation speed and quality and speed up the process of China's railway "going global".

3 Research and analysis

Neural network machine translation (NMT) created by Google is one the best machine translation engines at present. It takes each sentence as an independent neuron, breaking the phrase-based translation barrier. In this chapter, some contents from the first two chapters of *China's High-Speed Railway* [16] is selected and processed by Google translate. The MT output is compared with the final version after PE word by word, and the typical MT error type of railway engineering translation is analyzed, and then corresponding strategies are put forward.

3.1 Classification of MT errors

Based on the related researches on error types reviewed above and the analysis of MT output of the except, 4 typical MT error types of railway engineering translation are concluded, namely, lexical, syntactic, discourse and other errors, and then each category is subdivided. Finally, the typical types of MT errors in railway engineering translation are demonstrated in table 1. Understanding common errors in MT can help editors quickly recognize errors so that the editors can quickly deal with them according to corresponding strategies.

Table 1. Categories of MT Errors(C-E) for Post Editing.

Lexical errors	Syntactic errors	Discourse Errors	Other Errors
Mistranslation	Inappropriate translation of passive voice	Lack of coherence	Punctuation problems
Omission	Mistranslation of complex sentences		
Verb collocation	Inappropriate word order		

The excerpt selected has a total of 3892 words in Chinese. Based on the above error types, there are 129 errors in MT output, and the frequency and ratio of four error types are showed in figure 1.

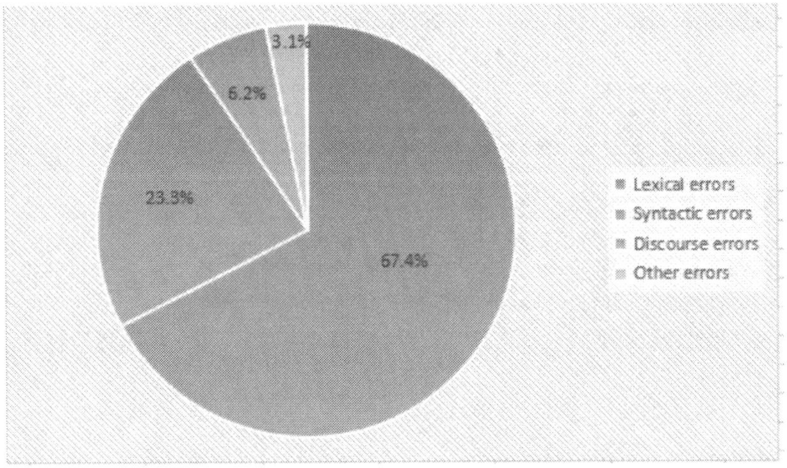

Fig. 1. The frequency and ratio of four error types of MT output of railway engineering text

According to statistics, there are 129 errors in the Chinese-English translation of 3892 words, including 87 lexical errors, accounting for 67.4% of the total errors, 30 syntactic errors, 23.3% of the total errors, 8 textual errors, 6.2% of the total errors, and 4 other errors, accounting for 3.1% of the total errors. From figure 1, it's clearly found that these four types of errors decrease in order of vocabulary, syntax, text and other aspects. Therefore, the editors need to pay attention to the types of errors in the process of PE, focus on the most frequent type of error and make corresponding modifications. Admittedly, the translation quality of Google translate is better than that of some traditional MT engines, but there still many errors in the MT output that need to be modified by human translators. In this chapter, some examples are taken from the MT output to analyze different types of errors in machine translation of railway engineering text and illustrate the corresponding strategies of post editing.

3.2 MT errors in railway engineering translation

3.2.1 Lexical errors

As analyzed above, it can be found that lexical errors appear most frequently in Google translate, accounting for 67.4% of all errors. The main errors are mistranslation, omission and inappropriate use of verb.

Mistranslation. We all know that in translation, the meaning of the word depends on the related context, but sometimes MT engine can't identify the subtle difference between synonyms and choose the inappropriate expression. In the analysis, it is found that mistranslation of nouns mainly lies in the mistranslation of proper nouns (terms) and common nouns. There are many special terms in the field of railway engineering translation. It's easy for Google engine to make mistranslations because of the lacking of professional knowledge.

Example 1: Including the addition of an open hole at the entrance of the tunnel when necessary.

In the above example, the underlined part is actually a terminology in railway engineering, but MT engine translates it literally into "open hole" which is obviously not correct. The correct translation should be "cut-and cover tunnel".

Example 2: The traction power supply system is the high-speed railway. Charger provides sufficient energy for high-speed trains.

Example 3: The main building is built to maintain the stability of the tunnel.

The mistranslation of common nouns is due to the inability of MT to distinguish synonyms according to different contexts. In example 2, the correct translation of the underlined part should be "electricity". Although "energy" has the meaning of the capacity of physical system to do work, but according to the context, it can be judged that the meaning of word in the original sentence refers to "electric power" required by train operation; and in example 4, "building" generally refers to formal buildings, mansions, etc., while "structure" focuses on the structural form. According to the context, the underlined part of original text refers to the composition of high-speed railway tunnel, so the use of "structure" is more appropriate here.

Omission. Omission means that some contents of source text is dropped off by MT engine, thus the information can't be completely conveyed. To deal with this problem, the translators need to supplement the information omitted in the target language to make the information complete.

Example 4: The Chinese standard EMU has a traction power of about 10,000 kilowatts at a speed of 350 kilometers per hour.

In this case, although the basic meaning of the original sentence has been translated, but a modifier is left out. The subject in this sentence has two attributives: "formation of 8 vehicles" and "with the speed of 350km/h", but it's obvious that the translation of the first attributive can't be found in MT output, so the information of the original text is not completely conveyed and the scope of China's standard EMU is enlarged. To solve the problem, the editors need to add the omitted information.

Verb Collocation. Word is used flexibly in English, which is mainly reflected in the selection of verb because subtle differences may produce a number of different verbs. When translating Chinese into English, the choice of verbs, especially synonym discrimination, is even not easy for human translation, let alone machine translation. For instance, to explain something, MT engine often translates it into "expand" or "elaborate". In fact, both the words have the meaning of "detailed description". However, the verb "elaborate" is used more in professional articles, so it should be used in the translation of scientific and technological texts. In a word, MT engine can't think like human translators and choose the correct verb according to different context.

Example 5: Its function is mainly to carry the lines on the bridge.

"Carry" and "bear" both have the meaning of "support and bear the weight or pressure", but "carry" lays particular emphasize on the moving items. In this example, what needs to be supported is a road, so the use of the verb "bear" is more appropriate here.

Example 6: Sufficient specific power to pull the train at high speed.

In example 6, "pull" means "the act of pulling" or "applying force to move something toward or with you", it often refers to the pulling of trigger or switch. Therefore, it's improper to use the word in this sentence. It's more appropriate to use the expression "enable trains to run".

3.2.2 Syntactic errors

According to statistics, the number of syntactic errors is second only to lexical errors, accounting for 23.6% of the total errors. Based on the study of typical errors types in MT output of professional automobile technology, LI Mei and ZHU Ximing [2] concluded that "among the syntactic errors, the order problem appears most frequently, followed by passive and infinitive problems, and the error rate is directly proportional to the length of the sentence".

Inappropriate Translation of Passive Voice. One major feature of English scientific and technical texts is the high frequency use of passive voice. However, in contrast with it, passive voice is seldom used in Chinese. On the contrary, the structure of active voice with omitted subjects is often used in Chinese. MT engine often translates the text in accordance with word order of the original Chinese texts, which is against the normal expression in English.

Example 7: Shanghai will use the Hongqiao hub as another new engine for Shanghai's economic development.

According to the source text, what the sentence wants to convey is that "Hongqiao Transportation is a new engine for Shanghai's development". However, MT engine directly translates the sentence in an active voice with "Shanghai" as its subject. But the subject of the verb "use" is usually human beings, "Shanghai" is a city, so the problem occurs. It's better to omit the original subject "Shanghai" and convert the voice into passive one, which has no effect on the meaning and more conform to common English practice.

Mistranslation of Complex Sentences. One of the characteristics of MT is: "for some simple sentences, there is no serious problems in MT output; but for some long sentences or sentences with a slightly complex structure, the quality of the MT output is not satisfactory, and sometimes it is even unreadable". To translate a long sentence, it is necessary to make a deep analysis of grammar, figure out the relationship between each part of the sentence, and then generate the translation according to the language rules of the target language. But it's difficult for MT engine to do this because machine translation tends to translate the sentence according to the grammar rules of the original text, connect each part into a sentence without any pause, thus resulting in the confusion of sentence structure and unclear meaning, which seriously affects the readability.

Example 8: The definition of my country's high-speed railway is: a newly-built passenger-dedicated railway with an EMU train of 250 km/h and above, and an initial operating speed of not less than 200 km/h.

In the above example, the definition of China's high-speed railway includes two parts: the newly-built multiple unit trains at the speeds of 250 km/h or above and the passenger dedicated railway lines in the early stage of operation carrying high-speed trains at the speeds of 200km/h or above. "The multiple unit trains" and "the passenger dedicated railway lines" have two modifiers respectively. However, MT engine mixes the two categories together and translates the sentence into "a newly-built passenger-dedicated railway with an EMU train", and the latter half of the MT output does not conform to the grammatical rules, which doesn't convey the original meaning accurately.

Inappropriate Word Order. Inappropriate word order refers to the phenomenon that very single word meaning is translated correctly by MT engine, but the order of the words is not conform to English practice, which will affect the readability of the translation.

Example 9: There is a good road, without a good car, nothing can be said about it.

Example 10: Without a good road, a car cannot drive, no matter how good it is.

In the above two examples, machine translation follows the order of the original text and translates the sentences word by word. However, the final translation does not conform to the common English practice. The structure of Chinese expression is relatively loose and short sentences are often used. However, English sentences are closely related according to the systematic grammatical structure, with many modifying and supplementary elements, and generally longer in length. According to the authentic expression of English, the sentences need to be reconnected.

3.2.3 Discourse errors

The application of translation strategies and methods of MT engine is limited to sentence level, and few translation methods are adopted at paragraph level or discourse level [18]. There is often no logical connection between sentences in MT output.

Lack of logic. The logical relationship in Chinese text is often hidden, while English is a language of"hypotaxis", and the logical relationship is more obvious. According to statistics, logical errors in this project account for 6.2% of the total errors, which is one the most common problems appear in MT output.

Example 11: My country has a vast territory, different climates from north to south, a huge road network, uneven population distribution and economic development in various regions, different requirements for long and short distance travel by passengers, and complex transportation needs.

In this case, machine translation roughly expresses the information of the original text. Logically, it seems that there is no problem. However, if the translator makes a deep analysis, it can be found that the logical coherence between sentences isn't conveyed in MT output. This sentence first introduces China's background knowledge to show the reason for "complex transportation demand". In fact, there is a causal relationship between them, which needs to be supplemented in post editing.

3.2.4 Other errors

Other error is mainly reflected in the misuse of punctuation, which appears less frequently than other error types, taking up 3.1% of all errors. In the output of machine translation, some punctuation marks may be missed or inaccurately used. In view of punctuation errors, the editors do not need to make special corrections, but only need to make the final check after revising the other three types of errors.

4 Strategies for PE

4.1 Lexical level

Choosing Correct Expression of Terminologies. When processing noun mistranslation in MT output, to deal with common nouns, translators need to compare several feasible words to define the connotation and denotation of their meanings accurately, and finally make judgments and choices of the proper expression. For terminologies, relying on the powerful online corpus, MT engines can quickly output the translation of terminologies, thus saving the translators a lot of time for verification. But in the terms of uncommon or special terminologies, the accuracy of machine translation can't live up to the translator's expectation. Therefore, in post editing, it is still necessary for translators to verify the translation of terminologies by checking the professional database or term base.

Choosing Correct Word Meaning According to the Contexts. In the process of post editing, based on the MT output, translators can quickly find words or sentences with inappropriate semantics, and then choose the closest lexical meaning in accordance with the context of the source language. In the cases of a word with many different meanings, translators should consult the online dictionary to determine the most reasonable explanation that can be integrated into the context. The choice of verbs, as important as that of nouns, plays a significant role in the accurate translation. Therefore, translators should pay attention to the synonym discrimination and collocation of verbs in the process of PE. By retrieving large English corpus, such as BNC, COCA, etc., the translators can quickly obtain the information of specific verb register, collocation and semantic prosody, so as to make a quick judgment.

4.2 Syntactic level

4.2.1 Transforming active voice into passive voice

In Chinese to English translation, translators can appropriately convert the active voice of the original text into the passive voice, especially when dealing with the sentence without subject. Over literal translation of MT engine often leads to the phenomenon of no subject in English, which will not only affect the semantic expression, but also bring misunderstanding to the readers. Therefore, experienced translators should consciously convert the sentence without subject into passive one, so as to avoid the phenomenon of no subject in English.

4.2.2 Segmenting and reorganizing the sentence structure

Each component of English sentence is closely related according to the systematic grammatical structure, with a lot of modifications and supplements, and relatively long in length; while in Chinese, the expression is relatively loose and short sentences are often used. When dealing with long sentences in Chinese, MT engine often follows the original text structure and outputs long sentences with mixed patterns. Therefore, when editing the MT output, it is necessary to analyze the sentence structure of the original text, segment and reorganize the sentence structure, and then output the correct expression.

4.2.3 Adjusting the word order

According to FENG Quangong and LI Jiawei [12], it is necessary to rearrange the word order to correct the errors caused by inappropriate word order which makes the translation obscure and can't be understand clearly. There are many obvious differences in word sequence between English and Chinese, and the machine mainly generates the translation according to the word order of the original text. Therefore, when editing MT output, it is necessary to adjust the word order reasonably according to the expression of English practice, so that the translation can convey the information accurately and make the expression more authentic.

4.3 Discourse level

4.3.1 Supplementing logical words

In view of the lack of logic, the translators only need to manifest the logical relationship in the target language in the process of PE. By means of adjusting the sentence structure or adding some logical words and conjunctions to make the logical relationship clear and translation more coherent and compact.

5 Conclusion

This paper selects railway engineering text for Chinese-English machine translation. Through manual analysis and comparison between MT output and MTPE output, four error types and their frequency distribution of MT output of railway engineering translation are discussed. Through detailed analysis of translation examples, this study puts forward the corresponding adjustment methods. Based on the research, there are some suggestions for improvement of the efficiency and quality of PE in railway engineering translation: first, proper pre-editing can reduce the errors produced by MT engine and improve the accuracy of machine translation, thus the workload of post editing can be reduced; second, the translator's own quality and ability is also one of the key factors to ensure the quality of PE, therefore, a qualified post editor of railway engineering translation should be equipped with not only bilingual competence, but also professional knowledge of railway engineering; besides, translators should improve their ability to obtain information in the new era; finally, practiced use of various computer-aided tools with high technology can greatly improve the efficiency of post editing.

The above research results can provide some references for MTPE practice of railway engineering translation, which is conducive to improving the efficiency of Chinese-English translation. However, the deficiency lies in that the samples analyzed in this paper are not supported by powerful theory, and the PE strategies provided can't be applied to all types of texts.

Acknowledgements

This research was financially supported by the Project "Study on Post-Editing of MT for Academic Works of Transportation Engineering" from Research Center for Applied Translation of Transportation and Engineering under Grant JD20017 and Jiangxi Provincial Humanity and Social Science Key Research Base, and Jiangxi Provincial Educational Reform Project "Integration Approaches of Internationalization and Localization in MTI Education" under Grant JXYJG-2018-090.

References

1. Q G Feng, H Y Zhang. To Train Post-Editors in the Background of the Global Language Services Industry. J. Foreign Language World, (01), 65-72, (2015)
2. M Li, X M Zhu. New Exploration of English-Chinese Machine Translation with Automatic Post Editing. J. Chinese Translators Journal, (04), 83-87, (2013)
3. Z S Li, J M Meng. A Concise Course of Machine Translation. M. Shanghai: Shanghai Foreign Language Education Press, (2009)
4. H H Lin. The Combination of Machine Translation and Human Intervention. J. Journal of Yibin University, (08), 114-116, (2009)
5. Q L Cui. Study on Post Editing of Machine Translation. J. Chinese Translators Journal, (06), 68-73, (2014)
6. J Hutchins. Has Machine Translation Improved. A. MT Summit IX: In Proceeding of the Ninth Machine Translation Summit.C. East Stroudsburg, PA: AMTA, (2003)
7. Y L Liu. Feasibility Analysis of "MT+PE" in Translation of Different Text Types. D. Shanghai International Studies University, (2020)
8. M S Plitt. A Productivity Test of Statistical Machine Translation PE in a Typical Localization Context. J. Prague Bulletin of Mathematical Linguistics, (93), 7-16, (2010)
9. A Guerberof. Productivity and Quality in MT Post-editing. C. In Proceedings of MT Summit Workshop on New Tools for Translators. A. (39), 137–144, (2009)
10. D Groves, D Schmidtke. Identification and Analysis of PE Patterns for MT. A. In Machine Translation Summit XII, (39), 429–436, (2009)
11. J Allen. Post-Editing. A. In H. Somers(ed.). Computers and Translation: A Translator's Guide. C. Amsterdam: John Benjamins Publishing, (2003)

12. Q G Feng, J W Li. Study on the Post Editing Model of News Translation. J. Technology Enhanced Foreign Language Education, (06),74-79, (2016)
13. Q L Cui, W Li. The Character of Error Types of Post-Editing: Perspective of Machine Translation Based on Scientific and Technological Materials. J. Chinese Translators Journal, (04), 19-22, (2015)
14. Q J Yan, W P Yan. Translation Methods of English for Science and Technology. M. Beijing: Metallurgical Industry Press, (1992)
15. H Kruger. Training Editors in Universities: Considerations, challenges and strategies. A. In John Kearns(ed.). Translator and Interpreter Training: Issues, Methods and Debates. C. London: Continuum International Publishing Group, (2008)
16. Z W Feng. Study on Machine Translation. M. Beijing: China Translation & Publishing Corporation, (2013)
17. K B Hu, Y Li. Research on the Characteristics of Machine Translation and Its Relationship with Human Translation. J. Chinese Translators Journal, (05), 10-14, (2016)

AUTHOR INDEX